MW00444701

# FINDING EQUILIBRIUM

# FINDING EQUILIBRIUM

## ARROW, DEBREU, MCKENZIE AND THE PROBLEM OF SCIENTIFIC CREDIT

TILL DÜPPE AND E. ROY WEINTRAUB

PRINCETON UNIVERSITY PRESS
PRINCETON AND OXFORD

Published by Princeton University Press, 41 William Street, Princeton, New Jersey 08540
In the United Kingdom: Princeton University Press, 6 Oxford Street,
Woodstock, Oxfordshire OX20 1TW

press.princeton.edu

Jacket Photographs: Detail of Kenneth Arrow courtesy of the National Science and Technology Medals Foundation. Photograph of Gerard Debreu courtesy of Biblioteca Universidad de Alcalá/D-Space/Creative Commons. Photograph of Lionel McKenzie courtesy of the University of Rochester. Background: Gerard Debreu papers, 1949–2001. Collection number BANC MSS 2006/218, Carton 8, research notes. Courtesy of The Bancroft Library, University of California, Berkeley.



Library of Congress Cataloging-in-Publication Data
Düppe, Till, 1977–
Finding equilibrium : Arrow, Debreu, McKenzie and the problem of scientific credit / Till Düppe and E. Roy Weintraub.
pages cm
Summary: "*Finding Equilibrium* explores the post–World War II transformation of economics by constructing a history of the proof of its central dogma—that a competitive market economy may possess a set of equilibrium prices. The model economy for which the theorem could be proved was mapped out in 1954 by Kenneth Arrow and Gerard Debreu collaboratively, and by Lionel McKenzie separately, and would become widely known as the "Arrow-Debreu Model." While Arrow and Debreu would later go on to win separate Nobel prizes in economics, McKenzie would never receive it. Till Düppe and E. Roy Weintraub explore the lives and work of these economists and the issues of scientific credit against the extraordinary backdrop of overlapping research communities and an economics discipline that was shifting dramatically to mathematical modes of expression. Based on recently opened archives, Finding Equilibrium shows the complex interplay between each man's personal life and work, and examines compelling ideas about scientific credit, publication, regard for different research institutions, and the awarding of Nobel prizes. Instead of asking whether recognition was rightly or wrongly given, and who were the heroes or villains, the book considers attitudes toward intellectual credit and strategies to gain it vis-à-vis the communities that grant it. Telling the story behind the proof of the central theorem in economics, *Finding Equilibrium* sheds light on the changing nature of the scientific community and the critical connections between the personal and public rewards of scientific work"—Provided by publisher.
Includes bibliographical references and index.
ISBN 978-0-691-15664-4 (hardback)
1. Equilibrium (Economics) I. Weintraub, E. Roy. II. Title.
HB145.D87 2014
339.5—dc23
2013050578

British Library Cataloging-in-Publication Data is available

This book has been composed in Minion Pro and Trade Gothic Lt Std.

Printed on acid-free paper. ∞

Printed in the United States of America

10 9 8 7 6 5 4 3 2 1

For Harald, Brigitte, Jens, and Benjamin (TD)

For Lauren, David, Molly, Alexa, Nico, Lucie, Henry, and RJ (ERW)

In some ideal sense, life philosophies, like economies, may be refined by successive adjustments through reflection, experience, and intellectual interaction with the past and the present until they come into an equilibrium independent of initial conditions. In fact, neither is ever independent of history.

**ARROW 1992, 43**

# CONTENTS

# PREFACE

> I think I finally understand the difference. We mathematical economists take a work, for example take Walras' *Elements*, and put it in boiling oil until all there is left is the skeleton, its essence or true character. You historians are instead interested in the skin and the flesh.
>
> **WERNER HILDENBRAND**

In the years following World War II, three mathematical economists—Kenneth Arrow, Gérard Debreu, and Lionel McKenzie—put, in Hildenbrand's words, economic theory in boiling oil and obtained one and the same skeleton: axiomatized general equilibrium theory. In their own histories, in their personal skin and flesh, they were as dissimilar as it was possible for economists to be in the mid-twentieth century.

Arrow grew up in immigrant Jewish New York. He was the precocious son of an established middle-class banker-lawyer who lost his job in the Depression and as a result the family found itself in greatly reduced circumstances during the 1930s. He was one of the Depression generation's bright kids educated at City College of New York, where he graduated at the head of his class in 1939. His academic career was superlative. As an undergraduate he took mathematics courses and was asked by Alfred Tarski to check the English translation of his German-language book on logic (Tarski 1941, xiv). Wanting to learn more about statistics because he thought he might find employment eventually as an actuary, Arrow ended up at Columbia studying with the eminent economist Harold Hotelling. Although his studies were interrupted by wartime service in the U.S. Army Weather Corps, after the war Arrow was recommended by Hotelling to the Cowles Commission in Chicago in 1948. He was fast-tracked to the highest levels of what was to become a new mathematical economics.

In contrast to the precocious urban Arrow, Debreu grew up in provincial Calais, was orphaned at an early age, and was raised by relatives and boarding schools. Late in high school his intellectual prowess became apparent as he finished second in a nationwide competition, the *concours général* in physics. After a strong pre-baccalaureate career, his teachers encouraged him to take the special examinations that would lead to entrance into one of the Grandes Écoles. Debreu was sent to Vichy in the south of France for that special examination preparation and after two years succeeded in securing admission in 1942 to L'École Normale Supérieure in Paris. His performance there was exemplary as he was taught by, and trained by, the very best mathematicians in France, particularly Henri Cartan among the emergent Bourbakiens. Following the Allied invasion in 1944, he served in the Free French Forces. After the war he sat his examinations and began thinking of ways to move away from mathematics and, serendipitously, made connection with Maurice Allais. Through Allais's intervention Debreu secured a Rockefeller grant to travel to the United States, where he eventually took up a research position at Cowles in Chicago. Debreu was as austere as Arrow was boisterous, as controlled as Arrow was exuberant.

Lionel McKenzie, in further contrast, was a well-mannered son of the interwar American rural South. Both his mother and father were well educated for that time and place, and McKenzie enjoyed a normal, intellectually successful but unaccelerated boyhood. With modest family financial resources in the Depression he went to Middle Georgia College and after two successful years there secured admission and a scholarship to Duke University. At Duke he performed at the highest level, receiving all As in honors courses concentrating in politics, philosophy, and economics, ultimately becoming the class valedictorian. In spring 1939 he won a Rhodes Scholarship that would have taken him to Oxford for two years had World War II not broken out in September 1939. He entered military service in 1941, and on demobilization he went to Oxford. Returning from Oxford after two years, and having begun a teaching career at Duke, he learned that he had failed to achieve the Oxford D.Phil. degree. He was no longer on a fast track.

Three men—a New Yorker, a provincial Frenchman, and a rural southerner—all joined the economics profession during the immediate postwar years. These three individuals, so dissimilar, were to have their names linked through two publications in the same journal, *Econometrica*, in 1954: Lionel McKenzie's article in the spring issue was called "On Equilibrium in Graham's Model of World Trade and Other Competitive Systems," and Kenneth Arrow and Gérard Debreu's joint article in the following issue was called "Existence of an Equilibrium for a Competitive Economy." Both articles established a fundamental theorem of modern economics: under clearly defined assumptions a competitive economy could have an equilibrium.

General equilibrium theory, in its simplest verbal form, describes nothing but the so-called circular flow of the economy. It concerns models in which, first, consumers attempt to maximize their gains from getting goods or services. They do this by increasing their purchases of the good until what they gain from the extra units are just balanced by what they have to give up to obtain the extra units. Likewise, individuals offer services (like labor) to firms by balancing the gains from offering the marginal unit of their services, their income, with the disutility of labor itself. This results in a theory of demand for goods and supply of productive factors. Second, producers similarly attempt to produce units of a good so that the cost of producing an extra unit is just balanced by the revenue received from selling that unit. In this way profits are maximized. Firms also hire factors of production up to the point that the cost of the additional hire is just balanced by the value to output that the additional hire would produce. In this way, costs are minimized and production factors are used efficiently. The firms thus generate supplies of goods and demands for factor services. This vision involves economic agents, be they households or firms, optimizing (doing as well as they can) subject to the relevant constraints. The decision problems are worked out in markets where prices represent the relative scarcity of all goods and services and thus guide the households and firms so their conflicting desires potentially can be reconciled.

General equilibrium theory is an overarching theoretical framework and, as such, it is neither testable nor falsifiable. It is instead a set of implicit rules or understandings for constructing particular theories that economists call "models"; those rules define the shared understandings of those who call themselves (mainstream) economists. The vision of the economy as an interdependent system can be traced far back in the history of economic thought, but its current role cannot be understood without the mathematical approval it earned by the contribution of our three protagonists: the proof of the existence of equilibrium. The problem is simply stated: can there indeed be a set of market prices such that the independent decisions that households and firms make on the basis of those prices clear all markets? The claim that "competitive market prices produce an efficient allocation of resources" is hollow if no such prices can exist.

Leon Walras in the 1870s was one of the first economists to pose the problem in this way. He provided an explicit argument to show that such an equilibrium was possible. For many decades economists lacked the means to assess the coherence of his proof, which had been based on a system of linear equations with a certain number of known and unknown variables. It was not until the late 1920s and early 1930s that Abraham Wald and other Viennese mathematicians tried to "prove" Walras's result (Weintraub 1983). The first mathematical proof of such a result using convexity arguments beyond calculus was done by John von Neumann in the 1930s. However, the 1930s proofs were constructed for model economies that were derived from assumptions that were difficult to interpret. It was not until the early 1950s that Arrow, Debreu, and McKenzie constructed models of a general competitive economy grounded in the independent decisions of households and firms and for those models established the existence of equilibrium by using the proof technique of the fixed-point theorem.

These proofs became highly influential. Economists gained confidence that the Walrasian vision of interdependent markets was indeed coherent. Economists knew that they could construct a model of a competitive private ownership economy in which equilibrium prices could be shown to exist. Constructing such a model, and establishing the possibility of equilibrium in that model, was a signal achievement in economic theory and a conspicuous element of what Roger Backhouse called the "transformation" of economics that occurred in the years before and particularly after World War II (Backhouse 1998). Rigorous mathematical proofs provided a different kind of closure to economic controversies. General equilibrium theory, as it was amenable to these new techniques of proof, helped establish the emergent disciplinary boundaries of economics.

There are many possible accounts of the emergence and subsequent success of mathematical general equilibrium theory in reshaping the profession of economics. But recognizing the diverse training and interests of the persons involved in the process of developing that theory, the historian of science is struck by one particular epistemic virtue that guided the new economics: the new modes of expression were *anonymous*. The abstract nature of the mathematical proof and of economic theory depersonalized the body of economic knowledge. It was now impermeable to the diverse personal visions and political inclinations that had long compromised the authority of economics. In the decades after World War II economics became less of a differentiated collection of Marxian, Keynesian, neoclassical, Marshallian, Ricardian, Institutionalist, Austrian (and so forth) economists. By "depersonalized body of knowledge" we see their work presented to a community that needs to know nothing about the authors to appraise their work. Keynesian theory on the other hand was a reflection of Keynes's own history and bears his personal stamp. Similarly Joseph Schumpeter's and Friedrich Hayek's personal visions shaped, and were intertwined with, their views on the development of capitalist economies. In contrast, Debreu could correctly say: "Even though a mathematical economist may write a great deal, it usually remains impossible to make, from his works, a reliable conjecture about his personality" (2001, 4).

Our narrative embraces the apparent paradox that the life stories of Arrow, Debreu, and McKenzie help us understand the impersonal character of economic knowledge in the postwar period. The anonymity of mathematics created complex relationships between the life and the work of our three protagonists. Our story explores the tension between the density and weight of their personal experiences and the aloofness and purity of the knowledge that they were creating. We see our job as historians as one of reconceptualizing the postwar mathematization of economics by repersonalizing it. Our story thus reconstructs the subjective nature of pursuing abstract knowledge in the social sciences. In what follows we examine the development of the new economics by following the career paths of three main characters of the general equilibrium story: Kenneth Arrow, Gérard Debreu, and Lionel McKenzie. We view the changes in economic theory through their life paths, their lived experiences, and their interaction in a small community that itself moved from the margin to the core of the discipline in the postwar years. The new mathematical culture in economics had personal ramifications well hidden by the newly accepted representation of economic theory as an abstract body of knowledge. We examine both the evolution of this economic theory through the experiences of those who brought it about and the development of the individuals' personae through examination of their work.

Past attempts to tell their stories were constrained by limitations naturally imposed on all historians of contemporary science: interests of vocal protagonist-subjects, incompleteness of the documentary record, unreliable memories of witnesses, and disagreements between scientist-subjects and historians over appropriate historiographic methods. But as the mid-twentieth century recedes in time, the historical past has become more accessible. Our own narrative will be based on newly open archives. This new and complex material occasions a comprehensive revision of the history of that period in economics.

What follows is in the tradition of work in the history of science that seeks to reveal the historical nature of modern knowledge (e.g., Porter 1996; Shapin 2010). As such, we will keep technical economic and mathematical analysis from overburdening the reader. It is not our task to deepen the reader's technical understanding of the new economic theory.[1] However, we do intend to make technical matters

1 Such has been earlier described by our protagonists themselves (Debreu 1984a; Arrow 1991; McKenzie 1999), as well as by historians of economics (Weintraub 1983; Ingrao and Israel 1990).

clear and obscure mathematical ideas comprehensible via the meaning these matters had for the work and careers of our agents. The "eureka moment" in a mathematical economist's intellectual life may be remote from common experience, but it is not inexplicable.

The continuing debate among scientists and historians of science about the connection between, and thus the importance or unimportance of, a scientist's biography and the scientist's work will not concern us here.[2] Our book is not a collective biography. It is rather a repersonalization of the theorems of Arrow, Debreu, and McKenzie constructed to gain a richer understanding of the postwar transformation of economic analysis. In connecting our protagonists' lives with their works we are not claiming that their lives determined their scientific work: like George Stigler, we believe it is nonsense to think that Leon Walras's toilet training caused him to construct the *tâtonnement* process. Nor do we claim that economists' personae are formed by the content of their work—a game theorist is no more a rational strategist in an intimate relationship than is a long-distance truck driver. Neither the work nor the personae are causal, one for the other. We will, however, explore their mutual or reciprocal stabilization.

The historian of science Thomas Söderqvist has recently characterized life-writing in science as an "edifying" genre (Söderqvist 1997, 2003). Like Söderqvist we insist that scientists' lives and their science are not separated by sharp boundaries. From life-writing in science we can learn to think of the meaning of the scientist's life for the science produced and, perhaps more important, of the meaning of the scientific work in the scientist's life. With Söderqvist, we view scientific practices thus as existential projects. Becoming a scientist is to recognize oneself, to forge an identity, as a "self-in-pursuit-of-knowledge." It is our task as historians to examine how this identification occurred or failed to occur. What coherence or dissonance was there between their lives and the science they brought about?

Our study will introduce a new *motif* to this genre of life-writing. Scientific credit is the screen onto which we will project the coher-

---

2 For a broad view of this topic in economics, see *Economists Lives: Biography and Autobiography in the History of Economics* (Weintraub and Forget 2007).

ence and dissonance between the scientists and their work. In a regime of impersonal knowledge, persons are not *naturally* attached to newly created knowledge. The process of crediting includes many practices by and through which this attachment occurs. Since this process involves both individual attitudes and strategies to gain credit as well as the many communities that grant credit, this new *motif* allows juxtaposing biographical and social accounts of science. Such repersonalization of the works "worth crediting" thus takes us beyond the sociology of the reward system in science as originally developed by Robert Merton.[3]

In the present-day organization of academic knowledge production, the problem of credit shapes a scientist's life; finding a job, publishing, creating protégés, and receiving science prizes involve credit. Generally speaking, there is a fundamental tension between credit given to individuals and the current ubiquity of research communities within which individuals' scientific work is produced and appreciated. Just as coauthorship is common in such a world, so is strategizing about credit: when one's coauthor has different interests, or if it happens that someone or another group is working on the same scientific problem, self-interests are in conflict and strategizing naturally emerges. We will see this clearly in the publication process involving our three protagonists.

Beyond the creation stage, credit is given by audiences at conferences but more importantly by editors and referees in both critical and editorial judgments leading to the publication of an article. In this setting priority in publication becomes salient. Subsequent to publication another phenomenon intrudes: credit accrues to scientists who create protégés in research groups or among students (Collins 1998, 24–28). Beyond one's own scientific offspring, the scholar gains credit from the work's influence in a larger community only indirectly related to the subfield in which it was birthed. This creates a certain distance between the context of origin and the influence of a work and encourages the reconstruction of the history from the

3 Our narrative instead draws from newer discussions of credit such as those by Gross (1998) and Biagioli and Galison (2003). Juxtaposing biographical and social accounts also makes our narrative complementary to the excellent study of Fourcade (2010), who compared the culture of twentieth-century economics in France, Britain, and the United States.

point of origin to the point of influence. This introduces yet another feature of the credit process: actors strategize on credit by creating their own historical narratives about their work. The history of science becomes part of the machinery that justifies and brings about credit. As late as 1988 Don Walker, then president of the History of Economics Society, proclaimed with no sense of irony that "[we historians] believe the writings of economists *should be judged* on the basis of whether or not they were original . . . whether or not [the work] was important" (1988, 101, emphasis added). It is such sensibility that generates the internalist survey-minded accounts that we are going to historicize in what follows. We are also aware that the same sensibility sustains the genre of life-writing in economics. Life-writing in science traditionally was a form of crediting a scientist, his genius as it were, and in writing a life the historian indirectly participated in the process of crediting. As Hankins pointed out, "The most obvious question to ask about biography and the reward system of science is whether biography itself is part of the reward system of science. We can easily answer that question in the affirmative. A biography honours a scientist" (2007, 93). This self-reflective element will appear and reappear in what follows.

The highest form of credit, however, is public symbolic credit in form of science prizes, the most famous of which are the Nobel Prizes (and the Fields Medal in Mathematics). For economics it is known officially as the Sveriges Riksbank Prize in Economic Sciences in Memory of Alfred Nobel and was created in 1969. Kenneth Arrow received this prize in 1972; Gérard Debreu received the same prize in 1983; Lionel McKenzie never received the Nobel Prize.

***

With our interest in the relationships among the *impersonality of mathematical expressions*, the *scientific persona*, and the *problem of scientific credit*, part 1 will examine the early careers of our protagonists in their attempts at forming scientific selves. That is, we try to understand the choices and needs that committed their intellectual lives to the emerging field of mathematical economics. What sense of meaning, what hopes led them into mathematical economics? Since this field was both at the margin of economics and at the margin of mathematics, what led them to accept the risks they took? What sense of personal meaning did they ascribe to their emerging professional identities? Such questions will concern us in chapters 1–3.

At the end of the war, our three protagonists were in their late twenties, a significant decade in life and memory. Each in their own way, more or less fortunately, took his place in the institutions of postwar science.[4] In part 2 we examine the contexts within which their contributions would later find an audience. Arrow's, Debreu's, and McKenzie's contributions would not have made sense, would have had little meaning, even a decade earlier. During the years just before and after World War II, the interaction between mathematics and economic theory was continuously renegotiated on various and unstable scientific and political grounds. The intricacy of the interrelated interests makes it hard to disentangle the skeins of economic theory, excitement about new mathematical tools, obligations toward political positions, and the influence of military funding. Yet as wide as were these ramifications of innovations in mathematical economics, so narrow was the community in which they took place. Those who pushed mathematical economics beyond calculus (which was the mathematics previously known among economists) were almost exclusively centered at the Cowles Commission for Research in Economics at the University of Chicago. Cowles scholars worked in a variety of research areas (Keynesian, Walrasian, statistical, theoretical, applied, and pure) with great enthusiasm about techniques and little thought about their influence on the profession at large. In chapter 4 we describe the sites within which the Cowles Commission was located and where each of our characters did the work that was to result in the existence proofs. Its intellectual atmosphere, particularly its openness to new technical solutions to old problems, facilitated acceptance of the existence proofs. The "air" that breathed life into the simultaneous discovery of the existence proofs became manifest at the conference on activity analysis held in June 1949 at the Cowles Commission in Chicago (chapter 5).

---

4 We are fully aware of our use of masculine pronouns in the narrative. As the photographs of the Duke and Rochester economics departments richly document, economics before the 1970s was almost entirely a male preserve.

The Cowles community was small. As Robert Solow later mused when reflecting on the difficulty in finding referees for journal papers in that period: "To whom could I have sent the [McKenzie] paper? I could have asked Paul [Samuelson], but he was always so busy that one hesitated to burden him. The community interested in and competent in those questions was trivially small. . . . It really was a tiny coterie" (letter to Weintraub, 2010). With respect to influence, hardly anyone at Cowles in those immediate postwar years expected that their group would one day move from the periphery into the center of economic research. "How small that group looks in retrospect," Debreu later exclaimed when giving a toast to Arrow's Nobel prize in 1972, "and how difficult it would have been to anticipate that several of your contributions would become standard parts of the graduate economic theory program" (GDP, 14).[5]

After part 1's presentation of Arrow's, Debreu's, and McKenzie's early careers and part 2's contextualization of the scientific world they inhabited in the late 1940s, we can better understand the specific roles they played in creating the existence proofs. In chapter 6, the heart of our historical analysis, we describe the connections among the three protagonists during their simultaneous work on an existence proof in general equilibrium theory. What were their respective strategies in dealing with the different values of the mathematics and the economics community? How did they respond to, and interact with, one another? How did they cooperate and strategize when dealing with a coauthor with different interests as well as another scholar who was writing simultaneously on the same problem?

In chapter 7 we examine the years following publication of the 1954 papers as those scientific results became important to the larger community of economists. The post-1954 experiences of our protagonists provide a window into the research practices in different mathematical economics communities that emerged during the cold war—at Stanford and Harvard with Kenneth Arrow, at Yale

5 Here and throughout the book, we will cite material obtained from one or another archive by (XYZP, ##), indicating the XYZ Papers, box or carton number. In cases where the material in the archive has been more finely processed, we will use the keyword of the folder name as well, and the date if available.

and Berkeley with Gérard Debreu, and at Duke and Rochester with Lionel McKenzie. As the values and interests of the Cowles Commission began to percolate through the economics profession via textbooks, dissertations, conferences, and new professional entrants, the meaning of those 1954 papers began to stabilize and shape the larger professional reputations of all three protagonists. Just as important, the three men came to understand their work somewhat differently one from another. Not only did their professional identities stabilize in that period, but their understandings and characterizations of their own contributions stabilized as well. Though Arrow and Debreu would later be credited for the same work, which eventually would be called the Arrow-Debreu model, they had different perceptions of their joint work. In addition, the community of economists and mathematical economists began representing and re-representing the Arrow, Debreu, and McKenzie results in new ways. We will thus approach the issue of credit by examining how the events of the 1950s set the path for the protagonists' respective careers during the years up to the Nobel Prize for Arrow in 1972 and for Debreu in 1983. We will also consider McKenzie's early professional marginalization despite his claims to have established priority in publication.

Though the Nobel Prizes appeared to settle issues of priority and credit, in fact they became more contentious. Employing distinctions about priority and credit first developed by the sociologist of science Robert Merton, in chapter 8 and the conclusion we will discuss issues of credit as the final point of the preceding genealogy of the postwar mathematization of economics. The meaning the three protagonists gave to symbolic credit, received or denied, makes manifest the coherence and dissonance between their lives and the science they brought about. It is for this reason that issues of priority and credit inform us about epistemic practices in a regime of anonymous knowledge like that which characterizes economics and other scientific disciplines today. In such a regime, credit becomes important as one of the few remaining aspects of science that links named individuals to new knowledge.

Mathematical work might be abstract, but we assert that it nonetheless expresses the life of the scientist performing that work. As Thomas Söderqvist (2003) has shown in the case of the immunologist Niels Jerne, science, as removed as it might be from art, and as much as it ignores the personal circumstances of the scientist-creator, in fact produces work that tells us about the life of the scientist. Science is not only a product of life but reflexively reveals aspects of autobiography. Personalizing the work of our protagonists thus raises an entirely new class of questions about the closure that economics achieved by adopting mathematical knowledge as an epistemic virtue. The rhetoric of science may modulate the scientist's voice, but as scientists tell us how to represent and intervene in the world, and in so doing show us the epistemic virtues to which they are committed, they are also presenting us with their personal way of experiencing the world. In what follows we seek to understand the peculiar human practice of economic theory by viewing it less as a way of representing the world than as a way of dealing with the world. And so we, historians of economics, will give voice to the mute experiences of those who dedicated their lives to the silence and discretion of reason so prized during the early cold war.

# CHRONOLOGY

| | |
|---|---|
| 1919 | **McKenzie:** January 26: Born in Montezuma, Georgia |
| 1921 | **Arrow:** August 23: Born in New York City<br>**Debreu:** July 4: Born in Calais |
| 1920s | **Arrow's** father loses his job at onset of U.S. Depression; family faces economic difficulties<br>**Debreu's** sister and mother die; father commits suicide; Debreu in boarding school |
| 1930s | **Arrow:** 1936: Accepted at City College of New York, major in mathematics. 1937: Internship as actuary. 1938: Proofreader for Tarski. 1939: Gold Pell Medal for highest average in all studies. 1940: Graduates with Praeger Memorial Medal for highest average in senior year, Ward Medal for Logic, and Phi Beta Kappa<br>**Debreu:** 1939: Wins second prize in National *concours général* in physics (travels to West Africa); goes to preparatory classes in Ambert and Grenoble<br>**McKenzie:** 1937: After Middle Georgia College (1935–37), transfers to Duke University (1937–39); Phi Beta Kappa in junior year. 1939: Wins Rhodes Scholarship |
| 1940–45 | **Arrow:** Fall 1940–fall 1941: Graduate school at Columbia (mathematics), takes Harold Hotelling's class in mathematical economics, reads Hicks's *Value and Capital*; moves to economics Ph.D. program; denied actuarial student internship; master's thesis titled "Stochastic Processes." Winter 1941: Oral Ph.D. examinations in economics. October 1942: Aviation training program at NYU. September 1943–July 1945: Weather officer in Asheville, NC. July 1945–February 1946: At weather division headquarters of U.S. Army Air Force<br>**Debreu:** Fall 1941–spring 1944: École Normale Supérieure in Paris, taught by Henri Cartan. 1943: Reads Allais's *À la Recherché d'une Discipline Economique*. Fall 1944: Officer school in Algeria. May 1945: Six weeks in Germany. Fall 1945: Aggregation at École Normale<br>**McKenzie:** Fall 1939–41: Princeton Graduate College in economics; reads Hicks in Morgenstern's class; hears talk by von Neumann on growth. 1941–45: Economist at Office of Civilian Supply at the War Production Board, then enters military service in navy as a cable censor in Panama and New York |
| 1946–June 1949 | **Arrow:** April 1946: At Columbia searching for a Ph.D. topic with Abraham Wald. July 1946: Invited to Cowles Commission. February 1947: Receives first offer from Stanford. March 1947: Turned down for Guggenheim fellowship. October 1948: Assistant professor at Cowles and Chicago department; discovers simultaneous result by Duncan Black. Summer 1949: Proves "impossibility theorem" while at RAND<br>**Debreu:** 1946–49: CNRS Fellow with Maurice Allais<br>**McKenzie:** January–August 1946: Teaches at MIT, meets Samuelson. Fall 1946: Pursues D.Phil. at Oxford under Hicks. September 1948: Goes to Duke. November 1948: Is informed that he failed the D.Phil. |

| | |
|---|---|
| June 1949 | Activity analysis conference at Cowles |
| Fall 1949 | **Arrow** accepts Stanford offer as acting assistant professor<br>**Debreu:** One-year Rockefeller fellowship mostly at Harvard. Fall 1949: At Cowles for "several weeks"<br>**McKenzie** abandons idea of Princeton Ph.D. after Graham's death |
| Fall 1950 | **Arrow** presents "On Extensions of the Classical Theorems of Welfare Economics" at Berkeley<br>**Debreu** becomes research associate at Cowles; presents optimality proof at Harvard, "The Coefficient of Resource Utilization"; is internal referee of Arrow's welfare paper<br>**McKenzie** goes to Cowles for the academic year; writes paper titled "Specialization in Graham's Model of World Trade" in Koopmans's class; meets Debreu |
| Fall 1951 | **Arrow** writes a "technical report" titled "On the Existence of Solutions to the Equations of General Equilibrium under Conditions of Perfect Competition" and leaves for a SSRC research visit to Europe<br>**Debreu** learns of von Neumann's proof from Slater; reads Kakutani and Wald in the English translation that contains no proofs; works on existence<br>**McKenzie** after returning to Duke begins to work on an existence proof in the Graham model; reads Wald's *Ergebnisse* papers and Morton Slater's notes on fixed-point theorems |
| Spring 1952 | **Debreu** reads Arrow's technical report and contacts him with comments, Arrow suggests collaboration; negotiations between Arrow and Debreu about which fixed-point theorem to use (Eilenberg-Montgomery versus Kakutani). May 29: Debreu sends paper "An Existence Theorem on Social Equilibrium" to John von Neumann to submit to the *Proceedings of the National Academy of Sciences* |
| Fall 1952 | **Debreu** invited to present his paper with Arrow, "On the Existence of a Competitive Equilibrium," for the December Econometric Society meeting in Chicago<br>**McKenzie** submits abstract of his paper "On Existence of an Equilibrium in Graham's Model of World Trade and Other Competitive Systems" for same meeting in Chicago |
| December 1952 | **Debreu** visits Arrow in Stanford to work on joint paper. December 27: Presents the Arrow-Debreu existence proof to the Econometric Society meeting<br>**McKenzie:** December 29: Presents his existence proof to the Econometric Society meeting; Debreu and McKenzie have different memories about this meeting of Debreu and McKenzie; Debreu does not inform Arrow about McKenzie's paper |
| Spring 1953 | **Arrow and Debreu** work to finish their paper; Arrow drafts "Introduction" and "Historical Note"; they negotiate about the axiomatic form<br>**McKenzie:** March 1953: Submits his paper to *Econometrica*; Strotz sends it to Nash and Hurwicz to referee. April 9: Submits to Solow an extension of his existence proof for the summer Econometric Society meeting in Kingston, Rhode Island. May 1: Solow acknowledges McKenzie's letter and notes the similarity between his paper and that of Arrow and Debreu |
| Summer 1953 | **Arrow and Debreu:** June 9: Paper is received by Strotz at *Econometrica*; Strotz sends it to Georgescu-Roegen, associate editor, to find referees, who forwards it to Baumol and Phipps<br>**McKenzie:** June 9: Strotz tells McKenzie he is re-prodding the referees. June 23: Strotz admits to McKenzie that there were failures in the referee process for his paper<br>**Arrow and Debreu:** July 17: Baumol sends his referee report on Arrow-Debreu to Georgescu-Roegen<br>**McKenzie:** July: *Econometrica* publishes the full abstract of the McKenzie paper from Chicago and says there was no abstract available for the Arrow-Debreu paper. August 6: Strotz tells McKenzie that the referees for his paper have been removed and begins looking for new referees. August 17: Hurwicz, original referee for the McKenzie paper, delivers his report and Strotz says he already has a second referee in place |

| | |
|---|---|
| Fall 1953 | **Arrow and Debreu:** Late September or first week in October: Phipps sends his report to Georgescu-Roegen rejecting the Arrow-Debreu paper<br>**McKenzie:** September 3: McKenzie gives his second equilibrium paper in Kingston to the summer Econometric Society meeting (no abstracts are published of that presentation)<br>**Debreu:** October 5: Debreu gets a request from Solow to referee the McKenzie article<br>**Arrow and Debreu:** October 8: Georgescu-Roegen tells Strotz that the Phipps report is worthless and that after receiving it he had done his own report on the paper, which he sends<br>**McKenzie:** December 14: McKenzie learns that the two reports on the paper are in and positive and that the paper will be likely accepted<br>**Debreu:** December 17: Debreu sends his final referee report about McKenzie to Strotz |
| 1954 | **Arrow and Debreu:** July: Paper appears in *Econometrica*<br>**McKenzie:** April: McKenzie's paper appears in *Econometrica* |
| Spring 1955 | **Arrow** rejects Nikaido's existence paper submitted to *Econometrica* as being no real generalization of the Arrow-Debreu proof<br>**Debreu:** After failed attempt at proving existence without fixed-point theorem, Debreu completes a full draft of the existence proof as it would appear in "Market Equilibrium" in 1956 and the *Theory of Value* in 1959 |
| Fall 1955 | Cowles Commission moves from Chicago to Yale. David Gale's "The Law of Supply and Demand" uses the same generalization as Debreu's and is published in *Mathematica Scandinavica* |
| 1956 | **Arrow** publishes first paper on stability<br>**Debreu:** June: Nikaido's "On the Classical Multilateral Exchange Problem" is published in *Metroeconomica*. Debreu's "Market Equilibrium" is published<br>**McKenzie:** Fall: McKenzie accepts offer by Rochester to become head of a new economics department and graduate program |
| 1961 | **Debreu** accepts offer from and moves to Berkeley after being denied tenure at Yale |
| 1967 | **Arrow** moves to Harvard |
| 1972 | **Arrow:** Nobel Prize to Kenneth Arrow jointly with John Hicks |
| 1974 | **Debreu** proves structural indeterminacy theorems after Sonnenschein and Mantel |
| 1980 | **Arrow** returns to Stanford from Harvard |
| 1983 | **Weintraub:** Publishes "On the Existence of a Competitive Equilibrium: 1930–1954" in *Journal of Economic Literature*<br>**Debreu:** October: Nobel Prize to Gérard Debreu |
| 2004 | **Debreu:** December 31: Debreu dies |
| 2010 | **McKenzie:** October 12: McKenzie dies |

# PART I
## PEOPLE

Without my work in natural science I should never have known human beings as they really are. In no other activity can one come so close to direct participation and clear thought, or realize fully the errors of the senses, the mistakes of the intellect, the weaknesses and greatnesses of human character.

**GOETHE [1836] 1998**

# CHAPTER 1
## ARROW'S AMBITIONS

Kenneth Arrow was never shy about engaging his past. In contrast to our other two protagonists, he gave a large number of interviews and on various occasions written sketches of different portions of his life and the development of his interests. Likely his openness to interviewers and biographers is the result of his ebullience and his lifelong interest in thinking about how ideas develop and how individuals' natures are formed. At the same time Arrow was hesitant about claiming the last word about his past. When asked to write about his "life philosophy" he began, "it is part of my life philosophy that no life can ever be examined fully and that attempts to do so are never free of self-deception. . . . Like the state in which I live, we plan and build on ground that may open beneath us" (1992, 42). Accordingly, the historian who undertakes to add another account to Arrow's self-accounts will wish neither to repeat them nor to reconstruct a self that is hidden by these self-deceptions. Thus we ask: Where did Arrow's intellectual ambitions come from?

### A Precocious Boyhood in New York (1921–40)

The first ground that opened beneath Arrow during his childhood was the Great Depression. He was born on August 23, 1921. Both of his parents had been born overseas and came to the United States as infants. They were both successful academically, as his mother graduated from high school and his father from college, not a usual event for immigrants in the 1900–1914 period. His father Harry's "family was very poor, [his] mother's hardworking and moderately successful shopkeepers" (Arrow, in Breit and Spencer 1995, 44). Arrow recalled that his father had some business successes fairly young and earned a law degree; he worked for a bank and as a result the family was fairly prosperous through the 1920s. With the Depression, his father lost his regular job and the family often had to sell household belongings in order to have money for food and rent and clothes. His father managed to do contract work for various legal firms from time to time in those years, but it wasn't until the end of the 1930s that the family began to reestablish itself economically.

Arrow was precocious. It was not simply that his school academic record was very strong, but he read extensively and deeply outside the school curriculum. He recalled that he read Bertrand Russell's *Introduction to Mathematical Philosophy* (1919) and other demanding books in philosophy, literature, and the sciences unrelated to his high school programs. On graduation he applied to Columbia University even though he was quite young (fifteen) compared to those who might have been his classmates. In an admission interview he asked the counselor about meeting the deadlines for financial aid decisions since he needed a scholarship in order to attend. The interviewer replied by telling him that he needn't bother about financial aid since he was not going to be admitted. In fact he was admitted, but the interviewer's comment had the effect of delaying the family's completing the scholarship application until after they heard about admission. By the time they realized what had happened it was too late for Arrow to apply for a scholarship. Many decades later Arrow discovered that his interviewer had been described by an historian as one of the most egregious anti-Semites in all of Ivy League education. Columbia, while not nearly as exclusionary with respect to Jews as were Harvard, Yale, and Princeton, had a *numerus clausus* (i.e., quota) arrangement to limit admissions of children of immigrant Jews living in the New York metropolitan region.

Without financial aid and with his family's own finances limited, Arrow applied to, and was accepted into, the City College of New York (CCNY). At that time, City College had free tuition for residents of the city, the result of earlier agreements and understandings that the prosperity of the city depended upon the education of its youth independent of their financial means. Admission was strictly by merit and Arrow was certainly meritorious. Moreover, CCNY was a commuter school so Arrow could live at home. CCNY was also, for the students especially, a particularly political cauldron. As the late Irving Kristol, the neoconservative editor and writer, recalled his student years at the City College:

> Every alcove [of the City College lunchroom] had its own identity, there was the jock alcove . . . there were alcoves for ROTC people—I don't think I ever met one—and then there was the Catholic alcove, the Newman Club. There was even, I am told, a Young Republican Club, but I don't think I ever met anyone who belonged to that club and maybe they didn't exist. But pretty much our life in City College was concentrated between alcove one and alcove two, the anti-Stalinist left and the Stalinist left. And that was our world, at least our intellectual universe. (Kristol, in Dorman 1998, 46–47)

Arrow was not so politically engaged as a student but was interested in many different subjects, and early on he decided that he would major in mathematics with the long-term objective of becoming a high school mathematics teacher: "I was concerned about getting a job. I didn't look beyond college very much at that point. All I wanted was security" (Arrow, in Horn 2009, 63). In that Depression period, secure civil service employment in the New York City public schools seemed reasonable to both him and his family. As a result, in his undergraduate program he took not only a lot of mathematics courses but also courses in education, and he did student intern teaching as well. In Arrow's case, that teaching consisted of conducting preparation classes for high school students who wished to overcome their initial failure on the New York State Regents exam through a retest process. He recalled that his students were the most motivated he had ever come across: "It was the biggest teaching success of my entire life" (ibid., 64). He loved teaching but a difficulty emerged: in 1932 the education administrators had constructed a list of qualified teachers from whom future recruits to the teaching profession would be drawn. The idea was that as teachers left the schools, those at the top of the list would be hired. But during the Depression teachers were not resigning to take other jobs. No new names, like that of Kenneth Arrow, could be added to the list until that preexisting pool of candidates had been drawn down. At least a year before his graduation, Arrow realized that he would not be able to find a job teaching high school mathematics. What was he to do?

In the summer after his junior year Arrow found work as an actuarial intern, even though he realized that if he wished to pursue this professional course after graduation, he would need to learn more statistics. Arrow thus took a course in statistics in the mathematics department, but the instructor apparently knew no statistics. However, one of the recommended books on the syllabus contained a large number of references to recent first-rate work in the field and so Arrow embarked on an independent reading program to become conversant with it all. He also did some other college work that would become important for his later success. In an interview with Jerry Kelly, Arrow spoke of how he encountered ideas in mathematical logic even though it wasn't part of any of his courses: he was "fascinated and used to aggravate my professors by writing out proofs in a very strictly logical form, avoiding words as much as possible and things of that kind" (Kelly 1987, 44). In his last term at CCNY he took a course titled "Logical Relations" with Alfred Tarski. As a result of his performance in that class, Tarski asked him to read and make necessary changes in the galley proofs of the English translation (from the German) of his textbook, *Introduction to Logic* (1941).[1]

Arrow graduated magna cum laude in June 1940, having won the Gold Pell Medal in his junior year for having the highest average in all his studies and the Praeger Memorial Medal for having the highest average in his senior year, as well as the Ward Medal for Logic. He was also elected to Phi Beta Kappa. He was not yet nineteen years old.

---

1 The translator, the philosopher Olaf Helmer (who will appear later in this chapter at RAND), was not a native English speaker and Tarski suspected that Helmer's translation might have been infelicitous.

## Columbia, 1940–42

"By the time I graduated, in 1940, the job situation was not very good. So when I asked myself in my senior year, what do I do next, I thought—well, why not go to graduate school?" (Arrow, in Horn 2009, 64). The only place to learn advanced statistics in the New York City area was at Columbia University. There it was taught by Harold Hotelling, who had succeeded Henry Ludwell Moore. But Hotelling, as Moore before him, was not in the mathematics department but in the economics department. Since Arrow had no interest in economics, he decided that he could take Hotelling's courses as electives if he were at Columbia studying mathematics. Arrow was able, with $400 given him by his father (who had successfully borrowed it since he "knew somebody who knew somebody who was rather well off" [ibid.]), to accept Columbia's offer of admission to the graduate program in mathematics for fall 1940: "I went to Columbia because . . . well there were several problems. One was that we were extremely poor and the question of going anywhere depended on resources. Columbia had the great advantage, of course, that I could live at home, which wasn't true anywhere else. I didn't get any financial support for my first year, none at all" (Arrow, in Kelly 1987, 45). To earn some money to help the family, Arrow got a summer job after his college graduation "as an actuarial clerk. That meant doing some elementary computations, calculating premiums . . . I was very fast, I picked it up immediately. I got paid 20 dollars a week" (Arrow, in Horn 2009, 68). Beginning his graduate studies with the plan to become an actuary, he soon was taken by, and "bought" by, mathematical economics:

> I had no interest in economics. I was in the mathematics department, taking courses like functions of a real variable, and I was going to take courses from Hotelling. In the first term he happened to give a course in mathematical economics. So out of curiosity I took this and got completely transformed. The course to an extent revolved around Hotelling's own papers. . . . Anyway, then I switched to economics from mathematics. I had gone to Hotelling asking for a letter of recommendation for a

> fellowship in the mathematics department [for my second year] and he said that, "Well, I'm sure I don't have any influence in the mathematics department, but if you should enroll in economics, I've found in the past that they are willing to give one of my students a fellowship." I was bought. Incidentally, I impressed him on about the second day of the class . . . he said he was puzzled by the fact that he had never been able to produce an example of Edgeworth's paradox with linear demand functions. So I sat down and wrote out the conditions for linear demand functions to yield the paradox; these conditions were certain inequalities on the coefficients and the inequalities were inconsistent. So I came to him the next day and showed it to him. Really it was just a few lines, but from that point on he was really impressed with me. . . . Anyway, I enrolled in economics. (Arrow, in Kelly 1987, 45–47)

Hotelling's assistant at that time was Abraham Wald, who had ended up in that position through a grant from the Carnegie Foundation after a year at the Cowles Commission. We will later see the deep connection between Arrow's work with Debreu and work Wald had done in the early 1930s. It is thus worth pausing here to introduce this remarkable figure.

Wald had been a mathematics graduate student of Karl Menger's in Vienna in the late 1920s, but since he was a Romanian Ostjude he was precluded from finding a university faculty position in Austria. As a result Menger sent him to Karl Schlesinger, an economist-banker in Vienna who wished to be tutored in mathematics. It was as a result of that relationship, and Wald's participation in Menger's mathematical colloquium, that he wrote two papers on the existence of a competitive equilibrium, one using a model based on a system of exchange and the other based on a model of production and exchange; he published them both in the proceedings of Menger's colloquium (Wald 1934, 1935). Wald solved the equilibrium problem by cumbersome brute-force techniques. His approach employed a very strong assumption about household behavior, an assumption that assumed that there was only one consumer, which simplified the argument immensely. It would be one of the major accomplishments of our three protagonists to construct a proof of the existence of equilibrium that followed a more natural economic logic than did Wald's proof. This point is important, and we will return to it in a subsequent chapter.

Oskar Morgenstern, a member of the colloquium, subsequently hired Wald as a researcher at the Business Cycle Research Institute he directed in Vienna (one of the several Rockefeller institutes in Europe organized on this subject). There Wald became a mathematical statistician and wrote an important monograph on seasonal variation in time series. Following the Anschluss, and Schlesinger's suicide, and in the face of Morgenstern's earlier unwillingness to nominate Wald rather than others for a Rockefeller Fellowship to come to the United States, Wald had to escape from Vienna. As a Romanian citizen, however, he needed to get travel documents there, so he traveled first to Romania and then by boat to Cuba before he was able to enter the United States with some support from the Cowles Commission then in Colorado Springs.[2]

It was during his time at Columbia from 1940 to 1942 that Arrow first met Wald while taking Hotelling's class in mathematical economics: "[As] I began to know a little more economics, I was hit by the number of extremely original papers that Hotelling had written. . . . What [Hotelling] taught was essentially the theory of the firm and the theory of the consumer. . . . I was a complete master at bordered Hessians" (Arrow, in Horn 2009, 67), a matrix of partial derivatives that was the main tool used in solving the kinds of optimization problems that Hotelling's class addressed. Arrow still hoped to secure a full-time actuarial position, and so he applied in the spring of his first graduate year for a summer student position. But even as a mathematics graduate student he was found to be unqualified:

> April 14, 1941. Dear Mr. Arrow: We've now reviewed the papers in connection with all applicants for actuarial student positions this year, and find that we have 27 candidates who have

2 The earliest discussion of Wald's role in the existence proof saga appeared in Weintraub 1983. Robert Leonard's definitive work on Menger's colloquium, which included extensive new material on Wald, appeared in 2010. The Wald of our account here is thus the Wald given to us by Leonard.

> passed the mathematical test. As we propose to employ about 6 students this year, I'm afraid a lot of good men will be disappointed. I regret to state that you were not selected to fill one of the vacancies open at the present time. Mr. [XXX], Associate Actuary, The Prudential Company of America. (KJAP, Accession 2008–0037, Early Career)

In spring and summer 1941 he wrote his mathematics master's essay, titled "Stochastic Processes," a copy of which is preserved in the Arrow Papers at Duke University (KJAP, 28, Master's Thesis). Fully engaged with statistical work while finishing his mathematics master's degree in fall 1941, Arrow sailed through his courses in economics and reached the dissertation stage very quickly. He passed his oral Ph.D. examinations by December 1941 in economic theory, public finance, statistics, and business cycles, while being certified in economic history and mathematical economics. He won a University Fellowship for 1941–42 and a Lydig Fellowship for 1942 (which he would not take up until 1946). It was in this period that he read John Hicks's new (1939) book *Value and Capital* and realized that there was a way to think about economics in a systematic fashion: "You know, after reading through the mish mash like Marshall and things like that, suddenly there was this clear, well-organized view, you knew exactly what was happening. Just the sort of thing to appeal to me" (Arrow, in Kelly 1987, 47). This would be the entry point for Arrow's first attempt at writing a doctoral dissertation, which he hoped would take a Hicksian approach to some Marshallian production conundrums.

Even though Hotelling was Arrow's primary mentor, the dominant presence in the Columbia department was Wesley Clair Mitchell, who spent his time downtown at the National Bureau of Economic Research and so was generally unavailable in the department. His course on business cycles was data based, and Arrow recalled that he appreciated the statistical care with which matters were treated. He also had a course from Arthur Burns, who replaced Mitchell as a teacher for a period of time, and from A. G. Hart, who would eventually (postwar) serve as his thesis advisor: "The place was a little bit weird, even by the standards of the time, in the sense that it was very anti-neoclassical. One of the results of this mood was that there was not a course in price theory, at any level. . . . [Mitchell] said it was our duty to collect a lot of data. When you have collected enough data, then things will be [clear]" (Arrow, in Horn 2009, 69).

## Launching a Career (1946–50)

Following the U.S. declaration of war on December 8, 1941, Arrow, who was certain to be drafted, enlisted in the hope of securing an officer's commission in the U.S. Army Weather Corps where he believed he would have a chance to use his mathematical and statistical training. He was quickly approved to attend an aviation training program at New York University in October 1942, taking "active duty" breaks from classes for rifle drill, which he and his colleagues thought rather silly. Nevertheless, he came out of that program in September 1943 commissioned as a weather officer with the rank of second lieutenant and was assigned to a weather research facility in Asheville, North Carolina; in July 1945 he was transferred to the weather division headquarters of the Army Air Force. It was during that time in Asheville that he wrote a memorandum that later, in 1949, became his first professional paper ("On the Use of Winds in Flight Planning" in the *Journal of Meteorology*). That paper presented an algorithm for taking advantage of winds aloft to save fuel on North Atlantic air crossings, an idea that was not acted upon by the military at that time but became the canonical practice for North Atlantic flight paths in the postwar period. He separated from the service on February 14, 1946, with the rank of captain.

Returning to Columbia in April 1946, Arrow needed to think about employment. During that summer he was hired as an instructor in economics at City College, and in the fall Columbia gave him an assistantship in statistics. He also returned to the problem of finding a dissertation topic. His first thoughts were connected with a course he had taken with John Maurice Clark that had examined the new work of the 1930s by Joan Robinson and Edward H. Chamberlin on imperfect and monopolistic competition. But he had other ideas as well. His curriculum vitae at the time recorded that he expected to complete, by February 1947, a thesis titled "Stability of Equilibrium in a Certain Microeconomic System," to be supervised by Abraham Wald with prospective dissertation committee members Clark, Arthur Burns, and Robert Haig, an early influential analyst of the U.S.

federal income tax. Hotelling had already left Columbia for the opportunity to create the first U.S. department of statistics at the University of North Carolina at Chapel Hill. This thesis appeared to be an extension from exchange to production of Hicks's material on stability of markets that had appeared in his 1939 book.

Arrow made very little progress on that or any thesis and at a number of points was quite prepared to give it all up, thinking that he would never be able to accomplish anything original: "Everybody thought I was very good. I thought I was very good as a student, too, but I wasn't at all sure I was capable of original work" (Arrow, in Horn 2009, 72). His continued interest in becoming an actuary was in conflict with his interest in a career of research in mathematical economics. Two letters to him in this period bracketed the possibilities:

> July 10, 1946. Dear Mr. Arrow, both Professor Hotelling and Professor Wald have mentioned your name with strong recommendations in connection with our search for a mathematical statistician-economist to succeed Ted W. Anderson in work at Cowles Commission research program. Both have indicated that you might wish to complete your Ph.D. work before continuing with other research work. However this may be, we would like to make your acquaintance and exchange ideas. . . . Sincerely yours, Tjalling Koopmans, Cowles Commission. (KJAP, 22, Koopmans)

Tjalling Koopmans, who would become a central figure in bringing our three protagonists together, at the time of writing the letter was the fugleman of Jacob Marschak, then research director at the Cowles Commission. How would Arrow respond given that he still planned to become an actuary? A second letter written by Hotelling on August 14, 1946, cleared matters:

> Dear Mr. Arrow: A letter from [a Mr. Cody] of Equitable Life inquires about you, stating that you have applied for a position in the actuarial department. In reply I recommended you highly. However, I'm inclined to sympathize with certain misgivings expressed by Mr. Cody as to whether you would really be satisfied with the kind of job in which you would find

> yourself. Practical actuarial work usually involves only elementary mathematics and becomes routine at times. As I wrote Mr. Cody, I imagine that you, unlike other men of active mind, high ability and extensive preparation, would be content with routine work for a time but would eventually become dissatisfied unless given challenging problems and greater responsibilities. . . . I had thought that your fellowship still had some time to run and that you would not be looking for a job at this time; otherwise I would have put you in touch with various positions in university teaching and also in practical statistics that have been opening up at a great rate. If you will let me know about your plans and desires with respect to further study, a job in the near future, and ultimate ambitions, I may be able to help. There are several very attractive statistical positions open in universities and industries. (KJAP, 22, Hotelling)

From then on Arrow stayed on the academic track. Hotelling suggested to Arrow that he apply for a prestigious fellowship outside the graduate economics program. In his application for a Guggenheim fellowship dated October 12, 1946, Arrow indicated that, as of that date, he had completed all requirements for the Ph.D. in economics except for the dissertation. He described his proposed fellowship project as: "Construction and possible empirical verification of a theory of major economic magnitudes based on a theoretical analysis of economic behavior of individual units of decision (Firms and Households); and the application of that theory to the formulation of ends and means in government policy" (KJAP, Accession 2008–0031, Guggenheim). This description of the project sounds much like the prospectus for the early eighteenth-century South Sea Company, one of whose investments was to be in "A company for carrying on an undertaking of great advantage, but nobody to know what it is" (Mackay 1980, 55), and it bore no coherent relationship to the thesis proposal that he hoped Wald would supervise. While the proposed study involved continuing his research for his doctoral dissertation at Columbia, Arrow also noted, "it is to be carried on using the facilities of the Cowles Commission for Research in Economics at the University of Chicago. This research group offers, I believe, a unique environment in which to carry on the project outlined"

(KJAP, Accession 2008–0031, Guggenheim). But by letter on March 24, 1947, he learned that he had not been awarded the fellowship.

The earlier letter from Koopmans had profound consequences for Arrow's career. At Koopmans's suggestion, Arrow went to talk with Koopmans at the American Statistical Society meeting at Cornell University: "There I got to know Koopmans. I knew that he had published a paper on a very interesting statistical point, and it was signed Pen[n] Mutual Life Insurance Company. So I asked him how he found working for an insurance company. And he just said, with his strong accent: 'Oh no, there is no music in it'. And the minute he said that, there was no further question in my mind about my joining an insurance company" (Arrow, in Horn 2009, 75).

After his meeting with Koopmans, he reconsidered taking a job at Cowles. That connection with Cowles would place him in the emerging postwar network of mathematical economists and economic statisticians. If he was good enough for Cowles, surely he was good. The proof was that even as he was about to be turned down for the Guggenheim, Stanford came calling.

> [February 5, 1947] Dear Mr. Arrow: Mr. Bowker has suggested to me the possibility that you might be interested in the vacancy in this department in the fields of advanced statistics and econometrics. The vacancy in question was created by the fact that Professor Allen Wallis left Stanford this year to accept an appointment at the University of Chicago. As a result of Mr. Bowker's suggestion, I have corresponded with Professor Wald and Professor Hotelling, and they both recommend you very warmly. I understand however that you're accepting a position with the Cowles Commission very shortly, and I realize that consequently you may not be free to consider the opportunity we have in mind in any event. It occurs to me however that you might be in a position to accept employment beginning, for example, January 1, 1948. . . . The position I have in mind would probably be one of rank of acting Assistant Professor, with the prospect of reappointment on a regular basis after one year if things worked out right." Signature B. F. Haley, executive head of the Department of Economics, Stanford University. (KJAP, Accession 2008–0031, Stanford)

Haley had understood correctly about Arrow's possible opportunities. But Arrow decided to stay on at Columbia for that year to work on his dissertation. In his published interview with Kelly, Arrow recalled that "at this time, I received an invitation to the Cowles Commission. At first I postponed a move because I was trying to finish my Hicksian dissertation before I went there, but I finally settled on finishing it there" (Arrow, in Kelly 1987, 49). In response to Kelly's questions about whether it wasn't unusual to leave graduate school before finishing a dissertation, Arrow replied, "You know, my knowledge of what was typical wasn't very good. The people I knew were at Columbia. I didn't know what was going on at Harvard or Chicago. It was really very provincial" (ibid.). As Kelly refocused his questions then about Arrow's invitation to Cowles, Arrow said,

> [T]hey came around and asked Wald and he recommended me. While he was primarily trained as a statistician, nevertheless he was interested in economics. There weren't many in that category and so they asked him. . . . [W]hat they really wanted me to do was to work on a statistical problem but it was a freewheeling place. At the moment, the emphasis was on the econometrics of large-scale models. So called "large" . . . three equations, five equations. Larry Klein ended up with a 20 equation model . . . but I was there to do anything I pleased and I was very obviously interested in theory. There was a feeling that theoretical foundations were also an essential part. Finishing my thesis could fit into this. . . . I really spent a year there not doing much of anything, to tell you the truth. I wrote a few tiny papers, none of which amounted to anything. I was a great contributor to discussions: argumentative, finding exceptions, errors and counterexamples. But I really felt very discouraged. (Arrow, in Kelly 1987, 50)

It was in the discussions at Cowles that Arrow developed a sense of the importance for scholarly productivity of being in an academic research community. He finally gave up his idea of pursuing a career outside academia.

> Dear Mr. Bennion: Regretfully, I feel that I cannot accept your offer of a position with the Standard Oil Company at this

> time. The major reason for my decision is the feeling that the progress in econometric model-building can best be served by continuing to work with a group whose interests and approaches coincide fairly closely with my own. Here, there is, in effect, a common language, which greatly facilitates the mutual criticism and exchange of ideas so important to scientific progress. Econometric research of the type I have been and will be doing is, like the work in the natural sciences, becoming more and more of a cooperative matter, requiring teams of individuals trained along similar lines. In the future it is hoped that the development will reach the stage where the results can be applied by individuals or businesses; but that time is not yet. (July 1947, KJAP, Accession 2008–0037)

The argument he was making—that economic research in the postwar period was becoming more "scientific"—was specific in its vision of that new economic science. It was not going to be moved ahead by isolated creative individuals writing books. Instead scientific economic knowledge would grow from work by research teams acting cooperatively. We have not found any other contemporaneous account of this change in economic research that lays out the prewar to postwar shift so clearly.

In October 1947, Arrow took up a one-year position as a research fellow at Cowles. After his fellowship was renewed for 1948–49, Chicago quickly realized Arrow's gifts, and on April 15, 1949, he received a letter from that university retroactively adjusting his appointment from October 1948 to a one-year position as an assistant professor of economics in the department of economics and as a research fellow (paid by the Rockefeller Foundation) in the Cowles Commission for Research in Economics through June 30, 1949, with a salary of $3,350 for the period. Moreover, for the period January 1, 1949, through May 31, 1949, he was to be paid for one-third time services on a contract between RAND and Cowles and for two-thirds time service by the university. He was twenty-seven years old, and his career was launched.

Nevertheless, Arrow was troubled by the gap between his ambitions and his performance up to that point. He was thrilled by the activity and energy of his Cowles colleagues but found he was unable to engage with a sustained project: "I felt very unhappy with my lack of progress. I was wondering whether I was really capable of having an original thought. I could critique, I was very smart, I could take the best economists and find what was wrong with what they were doing. But all this is part of being a bright student" (Arrow, in Horn 2009, 75). He was launched, yes, but personally he felt he had accomplished little or nothing.

We can pause here to note how curious this all appears in retrospect. In our modern world of economics, the path to a university career of scholarship and teaching requires extensive graduate training, competing with others for a very small number of jobs, and offering potential employers at least several papers in print or under revision for publication. It certainly requires a Ph.D. before one can begin a regular appointment. Arrow had none of these credentials. He was a young man of great promise, seen as having immense potential by both Harold Hotelling and Abraham Wald. He was energetic, enthusiastic, and verbally facile. But he was Jewish, which at that time made it impossible for him to find employment at schools like Harvard, Yale, and Princeton.[3] The Cowles Commission represented an academic path for refugee scholars and American Jews. Even though it did not have a sign saying "Exiles" at its front door, its motto "Science Is Measurement" attracted technically sophisticated scholars who were otherwise excluded from Harvard, Yale, and Princeton. It was at Cowles that Arrow's career was launched despite his lack of doctoral credentials and his being Jewish.

It was also there that love flourished. As Ross Starr wrote in *The New Palgrave*, "Jacob Marschak, the Cowles Commission Research Director, arranged for the Commission to administer the Sarah Frances Hutchinson Cowles Fellowship for Women pursuing quantitative

3 Wald at Columbia was a special case, but while postwar Columbia had eased its *numerus clauses*, it faced a threat from the New York City government of an investigation into its restrictions on hiring Jews (Karabel 2005, 210). The assimilation of European Jewish émigrés in the 1930s had not gone well even as non-Jews like Morgenstern and Karl Menger found a variety of opportunities (Weintraub 2014). In New York, the New School for Social Research was called the "University in Exile" and was home for a period to Marschak and Wald and others.

work in the social sciences. . . . The fellows were Sonia Adelson (subsequently married to Larry Klein) and Selma Schweitzer. Kenneth Arrow and Selma Schweitzer were married in 1947" (Starr 2008, section "Personal and Intellectual History," paragraph 7).

## Social Choice (1948–50)

Arrow's dissertation remained in limbo. He was unenthusiastic about the project, believing that it represented no original thinking or new and interesting analytic techniques. Without thesis supervisors pushing him to complete the exercise, he found the scintillating seminars and discussions at Cowles reason enough to put it off. His interests were eclectic, and his colleagues were doing fascinating work: "It was a terrifically exciting group, with Leonid Hurwicz, Lawrence Klein, Koopmans, Marschak, and some others" (Arrow, in Horn 2009, 75). Then, quite by accident, he started down a path that would bring real results in a fairly short period of time.

> Once, in lunch, we were talking about politics, left parties and right parties, and I remember drawing on a piece of paper the idea that a voter might have preferences over the parties. . . . So I wrote this thing down and started looking at the question of majorities. It's really hard to describe it. All I can say, is once you've seen it, it's obvious; it takes an hour or two. If you ask the question, the answer's fairly obvious. (Arrow, in Kelly 1987, 50)

The set of techniques that he knew to draw upon came from to his undergraduate course with Tarski on "logical relations." He began to realize that those relations were simply mathematical orderings and the apparatus of mathematical logic could be employed almost in full in understanding what was then being developed, at Cowles by the mathematicians Israel Herstein and John Milnor (and the next year by Gérard Debreu) and others, as utility representations of preference orderings.

> I spent a day or two working it up as a formal proof. And in my usual way, I sort of stalled about a month on writing it up for publication . . . but in a sense it didn't matter because within a month I picked up [the new issue of the *Journal of Political*

> *Economy*] and there's the paper by Duncan Black (1948) that had exactly that idea. (ibid.)

This was a shock. Arrow believed that this was the first time that he had had a very good, new idea. He wrote the paper expecting that it would at least salvage the Cowles year in which he had not produced anything of importance. He had even thought of using the idea as the major result of his dissertation.[4]

> This I still do not understand. I do believe in multiple discoveries, there are lots of them but usually it is because the ground has been prepared. Something happened, perhaps for some other reason someone developed something, and the next step was at least reasonable, it might not have been obvious, but at least reasonable. But what Black and I did could have been done 150 years earlier, there was no mathematical development, there was no intellectual development. Perhaps game theory in a very general way could be credited. . . . Why Black and I hit on this at about the same time, I really do not know. It was actually sheer chance. He could have done it a year earlier, I could have done it a year earlier. . . . I was really disappointed in not having priority since I had not published anything yet that was worthwhile. I thought at least to get some little note and then I was scooped. (Arrow, in Feiwel 1987, 192–93)

So at the end of that first year at Cowles, "I was still this brilliant person, very active in the seminars, but I really didn't get anything done. Another year [had] passed. Everybody expected me to amount to something. I was the only one who was doubtful about this" (Arrow, in Horn 2009, 76). Dissatisfied with his productivity and

4 Simply stated, the problem concerned a well-known paradox of voting. If there are three individuals, call them 1, 2, and 3, and three objects of choice, call them A, B, and C, suppose 1 ranks them in order as A then B then C. Suppose 2 ranks them as B then C then A. And suppose C ranks them as C then A then B. Majority voting on A versus B, B versus C, and C versus A will result in A ranking higher than B which ranks higher than C, which ranks higher than A! That's the paradox, attributed to the Marquis de Condorcet and called the Condorcet Voting Paradox. Black, and Arrow, proved that a simple assumption about preferences could rule out any possibility of the paradox's occurring.

newly frustrated by his having been "scooped" by Black, he decided to take advantage of an unforeseen opportunity.

> [T]hen that summer [of 1948] I went to the RAND Corporation—again through sheer accident. My wife, whom I met as a graduate student at Chicago, had previously worked in the Agriculture Department. She arrived there as a clerk and became a professional, a statistician. Her boss was a very distinguished mathematical statistician named M. A. Girshick. So I was friendly with Girshick who had gone to the RAND Corporation when it was started. . . . One of the things RAND was doing was inviting large numbers of visitors for the summer so Girshick urged me to come. Summer in Santa Monica didn't seem like a bad idea to me and it turned out to be far more intellectually exciting than anything I had planned because the halls were filled with people working on game theory. Everybody was fooling with zero-sum games, how to calculate them, the fundamental definition of the concepts; it was work at the conceptual level and at the technical level. (Arrow, in Kelly 1987, 51)

That 1948 summer at RAND was the start of his annus mirabilis. Game theory was much in the air, and he spent some of his time in Santa Monica learning about both the theory and its applications. One day, talking with Olaf Helmer (Tarski's translator) about games and politics, Helmer raised the following question: Suppose the players in a two-person game were the United States and the Soviet Union. Payoffs in that political game were to be based on the players' preferences. But the theory of preferences in economics and game theory was a theory of individuals' choices. What then could be the meaning of "the Soviet Union's preferences"? How did a group's preference connect to the preferences of the individuals who made up the group?

Arrow said that he could give an answer based on Abram Bergson's social welfare functions (1938) that ranked conceivable social states from lowest to highest, thus addressing issues of collective valuation of those social states. Helmer asked him to write it up as an expository piece. As Arrow began doing that, using the notion of "R" as an individual preference ranking of alternatives for a particular individual, he thought that "One natural method of taking a bunch of R's and putting them together would be by pairwise comparison by majority voting. And I already know that was going to lead to trouble! So I figured, well, majority voting was just one of a very large number of possibilities [of combining people's preferences], you just have to be more ingenious. I started to write possibilities down" (Arrow, in Kelly 1987, 53).

But as he worked on this, he began to realize that simple trial and error was not going to lead to any result, so he began to categorize the characteristics of various rules for aggregating individuals' orderings into a group ordering. He required that both kinds of orderings should have the same properties so that group orderings (or preferences) could function exactly as individual orderings did in game theory, which would allow one to talk intelligibly about, say, "American preferences" with respect to political outcomes. After a period of weeks he began to realize that none of the sensible aggregation rules was going to work and that only in the case of a dictatorship, where the dictator's ordering *was* the group ordering, would a group rule be unambiguously derived from individual orderings. It seemed that, without dictatorship, the problem would be impossible to solve.

> Finally, one night when I wasn't sleeping too well, I could see the whole proof, you know after playing around with it for a while. . . . I felt this was very exciting. I thought "This is a dissertation." You know, it's a funny thing. One of my problems had been feeling that one has to be serious and every time I'd thought about these voting questions they seemed like amusing diversions from the real gritty problem of developing a good descriptive theory. . . . But when I got the result, I felt it was significant. I really did. (Arrow, in Kelly 1987, 54)

The theorem was both unexpected and powerful: without dictatorship, no "sensible" method of aggregating individuals' preferences could lead to a coherent group preference. Put another way, it was impossible to aggregate individuals' preferences to generate the collective's preferences without violating some notion of "rational" collective preferences. Arrow presented the result that summer of 1948 at RAND and received a lot of useful suggestions from individuals there like Girshick, Abraham Kaplan, David Blackwell, and John McKinsey. On his return to Chicago that fall as a half-time assistant professor teaching statistics in the economics department, he presented the material in six discussion papers over a number of seminars (Hildreth 1986, 92). In the published monograph version (1951), he would thank Koopmans, Herbert Simon, Franco Modigliani, (statistician) T. W. Anderson, Milton Friedman, and (political scientist) David Easton for their helpful comments. He also presented the work at the December 1948 meeting of the Econometric Society in Cleveland.

> I remember Larry Klein was in the chair and Melvin Reder was reading another paper in the same session. My recollection was that there were 30–40 people in the room. I distinctly remember that in the audience was this contentious Canadian, David McCord Wright, who objected because among the objectives, I hadn't mentioned freedom as one of the essential values in social choice and apparently he went out of the room saying that Klein and Arrow were communists—this was quoted to me by at least Kenneth May who was present. . . . [Right] after the summer I developed this, on the way back to Chicago, I stopped at Stanford to be interviewed for a job. . . . Girshick had meanwhile moved to Stanford to contribute to starting a statistics department there. He was their star and he wanted me to join him. The economics department there had already in fact made me an offer a year earlier . . . I was appointed without a Ph.D. In fact, I got tenure without a Ph.D.[5] (Arrow, in Feiwel 1987, 56–57)

Through his work at Cowles and RAND, Arrow had quickly built up working relations with many members of the emerging mathematical economics community: "I already had one invitation from

5 This was technically correct since in those days, as Arrow recalled, one needed to have the thesis printed, and the degree could not be awarded until that event occurred. The work was done, and defended in January 1949, but he already had been hired by Stanford beginning in fall 1949, before the printing took place.

Stanford, but I wanted to get something done [before I accepted the Stanford offer]. . . . The editor of the *Journal of Political Economy* asked me to write an article for the journal. That was something I hadn't thought about. From that point on I became much more productive . . . I didn't go back to social choice. I didn't continue that line. I became much more creative" (Arrow, in Horn 2009, 77). He had tasted real professional success, and that success led him to other new ideas and new problems to solve. At the end of that Cowles year in June 1949, at age twenty-eight, he moved to Stanford where he would remain, with the exception of his midlife move to and then from Harvard, for the rest of his career.

# CHAPTER 2
## MCKENZIE'S FRUSTRATIONS

The 2009–10 telephone directory for the town of Montezuma, in Macon County, Georgia, lists nineteen McKenzies in a total population, from the 2010 census, of 3,460. These residents are nearly double the number reported in the 1920 census population of Montezuma of 1,827, the year after Lionel McKenzie was born there in 1919. The town, racially then as now approximately one-third Caucasian and two-thirds African American, was an agricultural town in an agricultural county in an agricultural state.

Apart from a few short notes on web pages, author information for books and articles, and some award citations, there is hardly any published biographical material on Lionel McKenzie except the short piece by Mitra and Nishimura (2009). He wrote only one autobiographical paper, which he read at Keio University in Japan in June 1998 on the occasion of his honorary degree: it was printed in *Keio Economic Studies* in 1999. There are only two other bits of autobiographical writing of which we are aware, single letters to two different individuals written in 1989 and 1990, copies of which he retained for his files (LWMP, Box 32). The first of these letters was to Morris Abram (July 28, 1989),[1] sent on the basis of someone's having sent McKenzie a copy of a very old *Atlanta Constitution* editorial that had

---

1 His law degree was from the University of Chicago following his undergraduate career at the University of Georgia; he earned the highest grade point average in the school's history at that time.

praised Abram's and McKenzie's appointments as Rhodes Scholars from Georgia in 1939.

McKenzie had read "with intense interest" (July 28, 1989, LWMP, Box 32) Abram's own then recent autobiography, *The Day Is Short* (1982), composed when Abram was quite ill with cancer. It was a tale of a boy from a Jewish family in the small town of Fitzgerald, Georgia, growing up in the hard segregationist years, who became an eminent civil rights activist and lawyer and later president of Brandeis University. His own background, so similar to McKenzie's in some ways, was nevertheless different in that he was Jewish at a time when there were few Jews in such towns in the South. Perhaps this is what led McKenzie to write to Abram as he did, sharing some of his own memories of place and family and Jewish presence in Montezuma in his early years. We reproduce most of it here, as it is in a voice and style not found in any of his published writings:

> [T]here were some Jewish families in Montezuma, from early times. The earliest of those I knew something of were the Happs. I had a picture of my father's baseball team at the turn of the century and the business manager, the only one in Mufti, was a Happ. Daddy was a pitcher. His crowning achievement was pitching two games in one day, winning both I believe. He left some notes on his life which are in a style of Mark Twain, I would say, not a copy, however, since he was not a reader. My Mother was a college student (one year at Wesleyan [in Macon, Georgia]—her father thought too much education was not good for women) and former schoolteacher. That no doubt explains her buying me the Book of Knowledge and Harvard Classics from door to door salesmen and starting me on my way. You may be interested to know that Mrs. Happ attended our Baptist church, not very orthodox I fear. My crowd also included a Jewish boy, Phil Brinen, who seemed to fit in very well but his family saved him by moving to Atlanta after high school. We lost track of him. I remember he once told me his father was fond of reading Tolstoy. That impressed me since Tolstoy was my favorite author at the time. We had a marvelous set of

> his works in the Carnegie Library beautifully printed. There was another family, the Hirschbergs, whom I knew in Oglethorpe (two miles away) because I collected for magazines from them. There was a son in that family who at this distance reminds me of you. Then there was Max Cohen or Bernd-Cohen, as he amended his name to make it more distinctive, who was a painter. I can recall my mother refusing to believe that Mary Mullino had allowed him to paint her in the nude, but I think she did. . . . Max got me an offer to attend the University of Pennsylvania, after Middle Georgia College, which would be paid for by John Lewis, a Philadelphia lawyer, while I boarded with Max who then lived outside of Philadelphia on the Main Line. Daddy and I were all for it but Mama objected. I think on religious grounds. That ended up in the expected way since I married a Jewish girl. It also lost me to physics and I'll bet a Nobel Prize because physics was my real love and I would probably have done well with it. (ibid.)

The narrative is charming, graced with affection for a time long past, with unclouded memories of growing up appreciated and comfortably matched in time and place. He was grounded in that southern town and subscribed nearly all his life to the town's paper, the *Citizen Georgian*.

Referring as he does to the books in his parents' house, and his mother's having been a schoolteacher with a college year at Wesleyan (in Macon, Georgia), the formation of his serious intellectual tastes is unsurprising (how many high schoolers read Tolstoy for pleasure?). Roman Pancs, who joined the Rochester faculty in 2008, recalled that only once did Lionel speak about his family of origin, saying that "he came from Georgia [and that] his grandfather was too old to fight in the civil war. The family had a series of late births—his mother gave birth when she was forty. His father was the oldest child" (Pancs, private communication).

He went to Middle Georgia College for the obvious reason—no "good school" would have given such a small-town southern boy a four-year full scholarship during the Great Depression. Nevertheless his record there qualified him for admission to one of those very good Ivy League universities. However, his real love was physics, and he would struggle with that field as a possible career choice through the early 1950s. This recollection appears in the second letter about his early years. He had read an article in the Montezuma newspaper about Langdon York, a star student at Macon County High School in Montezuma, the high school from which he himself had graduated. In a congratulatory note to Mr. York on March 11, 1990 (LWMP, Box 32), McKenzie wrote about his failed attempt to get into physics at West Point, the U.S. Military Academy. In order to be appointed to the academy, he needed to be nominated by his representative or senator. Georgia senator Walter George[2] interviewed him.

> I was struck by your plan to attend Middle Georgia College for two years before going to [Georgia] Tech because I attended Middle Georgia for two years long ago before going to Duke. I also noticed that you are interested in science and engineering which parallels my interest at your age. I had an appointment to West Point from Walter George but I allegedly failed the physical exam. In my opinion, I was rejected on ideological grounds. I answered a couple of questions of a political nature in a way that my interviewers did not like. Anyway, I later regarded that as a close call. After being at MGC for two years I was interested in a number of subjects including economics and physics. Possibly because I had read *Wealth of Nations* by Adam Smith from The Harvard Classics I chose economics when I entered Duke. I have often doubted that I made a wise decision. The upshot was that I tried to do economics as though it were physics which may not have been a good strategy. (ibid.)[3]

McKenzie here was projecting his own past on the young man's future and revealing himself as well. Pointing to his failed West Point interview, he did not tell of his (and Abram's) later successful interview for the Rhodes Scholarship that he won in his final year at Duke.

---

2 Walter F. George was a U.S. senator from Georgia from 1922 to 1957. See his official Senate biography at http://bioguide.congress.gov/scripts/biodisplay.pl?index=g000131.

3 The letter goes on to talk about McKenzie's connection with Richard Feynman, and he encloses a book by Feynman as a gift with his letter to that high school student.

Presumably progressive politics went further with the Rhodes committee than it did with the U.S. military in the 1930s. "Be true to your own vision" was his message to the young Mr. York. He also noted his having read Smith's *The Wealth of Nations* on his own, using his family's copy in the Harvard Classics, and that set his course toward economics. Wistfully, perhaps, he wondered what might have been different had he chosen physics instead; both here and in the letter to Abram he longed for a prize that eluded him because he had perhaps chosen unwisely. Economics might not have kept its promises that he felt from reading Adam Smith.

Again, the same sense of remorse was present in his biographical *Keio* piece when speaking about the choice of studying economics at Duke University:

> I think the book by Adam Smith . . . played a larger role in turning me to economics. It was also probably important that we were suffering from the Great Economic Depression then and my social studies course [at Middle Georgia College] based on ideas borrowed from a survey course at the University of Chicago emphasized the economic problems of my region and state. In any case the upshot was that when I transferred to Duke University . . . I chose to enter an honors course in philosophy, politics, and economics [PPE] modeled after the course of the same name at Oxford University.[4] I concentrated in economics. I should say that the strongest competition for economics in my life plans was physics to which I had been attracted by popular books on science by Eddington and Jeans.[5] I would be less than candid not to admit that I have often wondered whether my choice was a mistake. (McKenzie 1999, 1)

A retired scholar mentioning his early love for physics and speculating about an alternative life path is not "proof" of anything. But these three accounts can be read as reflecting his own hard-won un-

4 When Duke University created a PPE Certificate Program for its undergraduates around 2004, this was done with no awareness that the undergraduate college had ever had such a program.

5 Sir Arthur Eddington and Sir James Jeans were the founders, in the interwar years, of the new field of cosmology. Eddington wrote the immensely successful *The Expanding Universe* (1929) while Jeans sold many copies of his *The Mysterious Universe* (1930).

derstanding of his younger self or perhaps his belief about his own understanding. Late in his life he apparently felt the urge to make public his sense of remorse about his choice of field.

## Student Days: Duke and Princeton

Duke University, in the 1930s, was a school in the old southern tradition of fraternities, sororities, football, and classes that were quite strong in the liberal arts. It was a private university, affiliated with the Methodist Church, and supported generously by the Duke Endowment of Charlotte. With James B. "Buck" Duke's Indenture of Trust in 1923,[6] the university had the resources to build a new campus, moving two miles from the existing Trinity College (which was to become the Women's College), and thus in the 1930s it became a real university with professional schools of medicine, law, divinity, engineering, forestry, and nursing. Its neo-Gothic campus and English Perpendicular style chapel were unlike the archictecture of other southern schools, however, and its history of academic freedom and progressive idealism appeared to make it in but not comfortably of the South. Even as Duke's faculty grew dramatically over the 1930s and 1940s, the undergraduate students were a mix of mostly southerners with some northerners who had been attracted to a "southern" non–Ivy League school. At a time when serious education for women was still a curiosum in the South, well-established and well-off southern families often sent their young men north to the Ivy League, principally Princeton, while the brightest women students stayed in the region and went to either women's colleges or places like Duke, Tulane, Emory, and Vanderbilt. Nevertheless, the Duke Endowment exerted real influence over the Duke University Board of Trustees in order to keep Duke a southern university, not a national one. It was a football and fraternity/sorority culture, leavened by the usual dedicated teachers and scholars.

McKenzie's record at Duke University was exemplary. Winning honors and election to Phi Beta Kappa as a junior, he took economics,

---

6 James B. Duke was the founder of Duke Power Company and the son of Washington Duke, who created the American Tobacco Company and brought a small "Normal" school in Randolph County, a hundred miles away, to Durham to create Trinity College.

French, comparative government, economic thought, ethics, and social economics, receiving grades of A in every course. In his senior year he continued with economics and political science, doing these in special honors seminars while writing a thesis in his senior year, again receiving As in every course.[7] And so, in his final year, he applied for the prestigious Rhodes Scholarship to go to Oxford: "In my final year at Duke I won a Rhodes Scholarship from Georgia and I intended to pursue the same course as at Duke, popularly called PPE at Oxford. However this was not to be since the Second World War intervened" (McKenzie 1999, 1).

McKenzie graduated in June 1939 and was to sail for England in September. The outbreak of war on September 3 left him stymied. His financing for two years of study was in limbo. In normal times McKenzie, with the Oxford B.Litt. completed, could have returned to the graduate or professional school work that was to his taste, leading to a desirable scholarly research career. However, with the new school year starting in late September in the United States, he needed to move quickly to make new plans and to garner scholarship support. With his Duke record, and the Rhodes, his merit was undeniable, and his Duke professors were helpful. Frank de Vyver had recently arrived from Princeton to teach labor economics at Duke and it is probable that de Vyver, as well as Joseph Spengler and Calvin Bryce Hoover (who each had Princeton connections), encouraged McKenzie to go there. Princeton held a special place in the hearts of southerners compared to Harvard and Yale, which were seen as New England schools. Virginian Woodrow Wilson's Princeton presidency had made this connection visible. About Princeton itself, McKenzie recalled: "I applied to Princeton for support in their graduate program and entered the Princeton Graduate College in the fall of 1939 to work toward a Ph.D. in economics. I knew very little about either economics or the character of the Princeton department when I entered there" (McKenzie 1999, 1).

---

7 These grades suggest an astonishingly competent young scholar since As then were not what As have become today. For example, as late as 1955, the economics department at Duke required its faculty members to award no more than 11 percent As to seniors and to fail 1 percent, while only 10 percent of the freshman economics students could receive As and 10 percent had to receive Fs.

We can appreciate McKenzie's taste for understatement. As of September 3, 1939, he had not planned to become a graduate student in economics, nor had he any interest in spending his next two years anywhere but at Oxford. Making the best of the situation, he went on to recall that

> I found a number of quite stimulating professors in the program. I would in particular mention Frank Graham and Oskar Morgenstern. . . . Morgenstern taught me advanced economic theory where we read and criticized the new book *Value and Capital* by John Hicks. . . . I should mention that Morgenstern was not uncritical of Hicks's book. In particular he ridiculed Hicks's use of the equality of the number of equations and number of variables to conclude that an economic equilibrium existed. (McKenzie 1999, 1–2)

Indeed, Morgenstern wrote a scathing review of *Value and Capital* for the *Journal of Political Economy* in 1941. Some of that review was related to conversations Morgenstern had had with John von Neumann at that time as they were writing the *Theory of Games and Economic Behavior* (1944): in his diary Morgenstern reported von Neumann's contemptuous comment that Hicks's mathematics was that of the eighteenth century and was of no serious interest to anyone concerned with making economics a scientific subject (Mirowski 2002, 138–39n41).

At the same time that McKenzie was hearing about this criticism of Hicks's failure to establish the existence of a competitive equilibrium from Morgenstern at Princeton, Arrow was learning about it at Columbia in Hotelling's course on mathematical economics. During the war Arrow would think occasionally about Hicks's equilibrium theory and how he might try in his eventual dissertation to extend that system and analyze its properties. After he returned to Columbia in 1946 to write a dissertation, his reconnection with Wald rekindled his awareness of the difficulty of establishing an existence proof for a competitive economy. McKenzie, at Princeton, had a similar connection with Abraham Wald via Oskar Morgenstern, but it is unclear precisely when McKenzie became aware of Abraham Wald's work: He recalled that "Morgenstern had known Abraham Wald and John von Neumann in Vienna, and he was familiar with the papers on existence of equilibrium by Wald and von Neumann which were read to the colloquium by Karl Menger. However he did not give us references to these papers in the seminar" (McKenzie 1999, 2).

As Weintraub has shown (1983), the Mengerkreis, or Karl Menger's mathematical "circle" at the University of Vienna from the late 1920s until the Anschluss (Hitler's "annexation" of Austria) in 1938, was the site of the first modern mathematical treatments of economic equilibrium and growth.[8] Although most of the papers presented and published in the proceedings of that colloquium were in mathematics, Menger's own connection to economics, and his interest in a scientific economics, opposed the antiliberal and anti-Socialist views of Vienna economists like Othmar Spann (Caldwell 2004; Leonard 2010). Morgenstern and Wald were in the Mengerkreis (as was Kurt Gödel) and von Neumann was a regular visitor when he passed through Vienna. Once von Neumann and Gödel arrived at the Institute for Advanced Study in Princeton, interaction between leading mathematicians and economists was inhibited, as we will see in part 2, despite the collaboration of Morgenstern and von Neumann. McKenzie later recalled an economics department talk given by von Neumann (see chapter 6).

In addition to his classes with Morgenstern, McKenzie "also benefited from a course in mathematical economics led by our resident statistician, Atchison Duncan. . . . Duncan also encouraged me to read Wilson's *Advanced Calculus* which I did one summer despite the complete lack of preparation in calculus" (McKenzie 1999, 2). McKenzie's Princeton transcript identified him as a single white male who entered the graduate school in September 1939; he passed his general examination in spring of 1941, having earlier passed both French and German language requirements. In his two years of coursework, 1939–40 and 1940–41, he took classes in statistics and basic economics, including history of economic doctrines, but took additional courses from Frank Graham, Frank Kemmerer, Morgen-

8 Robert Leonard's (2010) award-wining study on von Neumann, Morgenstern, and game theory is the definitive scholarly treatment of this period and the community of Viennese economists and mathematicians.

stern, Duncan, and Friedrich Lutz, with his 1941–42 doctoral research work being supervised by Morgenstern. Graham was an international trade theorist whose model of world trade was to become McKenzie's vehicle for establishing the existence of a competitive equilibrium, while Kemmerer was the renowned "money doctor" whose missions in the 1920s and 1930s to various South American countries signaled their seriousness in confronting and fixing their failed monetary systems. After all his coursework was done, "I remained one further year at Princeton after completing my coursework and began a project with Morgenstern on futures markets, which was never completed" (ibid.).

## Wartime Service and (Finally) Oxford

McKenzie, who like Arrow entered graduate school in 1939, spent three years studying economics in the prewar period. He was of draft age, and the war would eventually, even if not immediately, require his active duty military service. For McKenzie that meant, at first, that he "spent about a year in Washington as a junior economist with the Office of Civilian Supply in the War Production Board. . . . There I was regarded as a theorist but I don't believe I made any significant contribution to either theory or the war effort." During that period in Washington he married Blanche Veron.[9] He soon joined the navy in a noncombatant role as a cable censor first in Panama and then in New York. Time in the navy, for McKenzie, allowed him to follow his intellectual passions: "When I was in Panama . . . I [did] what

9 There is almost no published material on Blanche Veron McKenzie, who predeceased her husband in 1999. Sketchy academic catalogues show that she graduated from Brooklyn College with an A.B. in 1941, which was the end of the second of McKenzie's three years at Princeton. We do not know if they knew one another while he was at Princeton, but circumstantial evidence suggests that they did, since for the next academic year, 1941–42, she received a resident scholarship from the department of economics and politics at Bryn Mawr College (in suburban Philadelphia) to participate in a college-funded research project. Train travel between Bryn Mawr and Princeton, then as now, was regular and quick. It would appear then that she left after that year to go to Washington, where she and Lionel married. In 1944–45, she is listed as a graduate student (presumably in economics) at Columbia University: this would be during the time that Lionel had returned to New York as a navy censor. She was apparently an intelligent and sympathetic woman, and her presence in the lives of women graduate economics students at Rochester was profound. Indeed, one of those successful students, Randi Novak, created an undergraduate scholarship in her honor at the University of California at Santa Cruz. The award's website mentions Blanche McKenzie's advanced degree, a master's degree certainly, in economics.

amounted to an undergraduate course of reading in mathematics and physics, as well as a fair amount of reading in both modern and ancient philosophy. . . . I was mustered out at the end of 1945 in response to a plea that I needed to undertake my Rhodes Scholarship" (McKenzie 1999, 2).

He did not undertake his scholarship right away. Out of the navy before the second semester of the academic year had begun, and with a wife and hopes for a family, he wanted to find a job but had only a master's degree and three years of graduate work in economics. But then he got very lucky. Demobilization had speeded up around Christmas 1945 as more shipping became available to bring troops home from overseas, so universities saw a very large increase in enrollments, particularly since many of the returning troops had deferred or suspended their college work during the war. With the memory of the 1930s Great Depression quite salient, this wave of students was especially interested in economics and business. Passage of the Servicemen's Readjustment Act of 1944 (known as the GI Bill) provided scholarship support for college work for returning war veterans, so universities needed to hire economics teachers quickly beginning in January 1946. MIT in particular faced these large enrollments of students wanting to prepare for careers in a more technical economics. Thus McKenzie

> obtained [a temporary] appointment in 1946 as an instructor at MIT, thanks to the recommendation of my former Princeton colleague Ansley Coale . . . this was after Paul Samuelson had become the star in residence publishing his *Foundations* and preparing his *Principles* text.[10]. . . I attended Samuelson's graduate class on economic theory but I remember little about the content. I do recall asking Samuelson in his office to explain his theory of revealed preference to me but I did not recall that he succeeded. My teaching at MIT was in industrial organization, where my expertise could be questioned, rather than in theory.

10 This recollection is certainly mistaken, since *Foundations* was not published until 1947. Samuelson remarks, in its preface, that it was written primarily in 1937, but it was not submitted as a dissertation until 1941. He wrote his preface to the published version in January 1945. *Principles* was not published until 1948, though drafts were used as early as 1946 in mimeo form at MIT.

> After two semesters [spring and summer] there I resigned to take up my deferred scholarship at Oxford. (ibid.)

This short period at MIT, where he became acquainted with Samuelson, would be extremely important as McKenzie's career unfolded. It would be Samuelson who would later help him as he sought employment after his time in Oxford.

In 1946 Oxford was neither a rich nor a comfortable university. The war had reduced its resources, and rationing of food, gasoline, and other amenities was severe.[11] As a married man, McKenzie could not live in the college but had to find lodgings in town. That added expense could be met with highly valued U.S. dollars since McKenzie had not only the Rhodes Scholarship but some accumulated savings as well as a grant from the Social Science Research Council (SSRC). Although he may have wished he could have taken up physics or philosophy in those Oxford years, he continued his program in economics, hoping to build on his three Princeton years to write a thesis for a D.Phil. degree that would suffice to secure a very good university position in the United States. Having already studied Hicks's *Value and Capital* at Princeton, he was pleased that Hicks was his D.Phil. program supervisor. The Oxford period allowed him to meet young economists in England like Ian Little, Paul Streeten, Jan van de Graff, and William Baumol. Oxford was not all economics, however.

> I spent very much, perhaps most, of my time at Oxford paying attention to subjects other than economics. In particular I attended several lecture series in philosophy with Gilbert Ryle and also Friedrich Waismann who had been a member of the Vienna Circle of philosophy in earlier years. . . . Of course, I attended Hicks' class which involved reading and discussing articles in the journal literature, especially recent ones. The project I was attempting with Hicks was an examination of the modern welfare economics to which Hicks had made contributions. Ian Little who had attended Hicks's class, along with Paul Streeten, was also writing on the subject and was my closest associate at

---

11 See the recent discussion of life as an American student in that period at Oxford by another "southern" Rhodes Scholar, also from Duke, the novelist Reynolds Price (2009).

> Oxford. As it happened the leading economics scholar at the LSE [London School of Economics], William Baumol, and the leading scholar at Cambridge, Jan van de Graaff, were writing on the same subject. Indeed we three gave the three presentations to the joint Oxford-Cambridge-LSE seminar in the academic year 1947–48. (McKenzie 1999, 3)

McKenzie had support for two years, 1946–47 and 1947–48. Thus he soon had to try to find a "real" job to return to the United States. Unlike Kenneth Arrow in 1948, he had few good options. He had kept in touch with his teachers at Duke, and with Samuelson, hoping they might help him find an academic job. On November 25, 1947, Samuelson replied to Vernon Mund of the University of Washington about McKenzie's suitability for a position there: "I formed a very favorable impression of McKenzie's promise as an economist. He is intelligent, with a creative interest in economic theory. . . . He was an excellent teacher here" (PASP, 52, McKenzie). And to answer the rather coded question always on the table in those McCarthyite years, he noted that "At no time did [McKenzie] impress me as being of a doctrinaire turn of mind, and so I should expect him to carry out careful and balanced investigations" (ibid.). But no job was offered. He needed to find one, though, as by 1948 he had not only a wife but a child as well.

Samuelson continued to work on McKenzie's behalf: on April 1, 1948, he wrote to Hicks and after passing along some professional news mentioned that "Lionel McKenzie has written me concerning possible job openings on his return to the United States next year. Would you give me your dispassionate opinion of his strengths and weaknesses so that I may be guided accordingly in recommending him for any openings that may develop here or elsewhere?" (PASP, 37, Hicks). Hicks's reply was informative: "When he started to work with me I was not very much impressed. I thought he was a bit of a dilettante. But for the last few months he has gotten down to write, and I am really very impressed by what I have seen of his thesis. . . . He attends my seminar and takes quite a leading part of the discussions. Really, on his present form, I think he ought to be backed" (ibid.).

A few months earlier (in late 1947) McKenzie must have written to Duke's Frank de Vyver seeking help about job prospects. In a letter from de Vyver to economics chairman Calvin B. Hoover dated January 15, 1948, de Vyver names three former students, McKenzie first among them, whom Hoover might wish to "recommend" if Hoover were to be asked for names of suitable candidates for academic jobs (CBHP, 43, McKenzie). De Vyver's letter was successful. Hoover wrote to the president of the College of Charleston on February 11, 1948, recommending McKenzie for a faculty position there (ibid.). But Hoover, seeking to increase the size and reach of the Duke economics department, and recognizing how Duke administrators looked with favor on Duke graduates, southern men, war veterans, and Rhodes Scholars, worked to bring McKenzie back to Duke.

He got Joseph Spengler to solicit a letter of reference for McKenzie from Hicks in support of McKenzie's suitability for a position at Duke. Hicks answered that "I am very favourably impressed by the work McKenzie has done under me, especially during the last six months when he has actually been writing his thesis, much of which I have seen. . . . I feel sure he would make a very good teacher" (March 17, 1948, CBHP, 43, McKenzie). It is interesting how Hicks's letters to Samuelson and Spengler differed. It is as if Hicks wanted McKenzie to have a better opportunity than he thought Duke could provide, since his letter to Samuelson was stronger than was the letter to Spengler.

In the first week of April 1948 Chairman Hoover concluded negotiations with Duke's Dean Wannamaker to make McKenzie an employment offer: "We are able to offer you the position of Assistant Professor at a salary of $4,000 for the nine months [to teach four courses each semester with at most three preparations].[12] This offer would be contingent on your receiving the Ph.D this year. If you were delayed in receiving your degree the salary would have to be somewhat lower with the understanding that it would reach $4,000 when you received the degree" (CBHP, 43, McKenzie). Thus on April 13

12 In fall 1948 he taught three sections of Principles I, and one of Principles II, with enrollments of 41, 43, 28, and 39, respectively.

McKenzie sent a handwritten note to Samuelson telling him that "I have decided to accept an Asst. Prof. at Duke for next year. The salary is quite good. I don't think I could improve it. And also it is my undergraduate university" (PASP, 52, McKenzie).

Perhaps McKenzie had been insufficiently committed to economics at that time, or perhaps he was more intrigued by physics and philosophy, or perhaps he saw in Oxford economics an intellectual backwater compared to Princeton and MIT, or perhaps his thesis supervision was inadequate. The result was unfortunate: his Oxford examining committee failed his D.Phil. thesis. After he had returned to the United States and begun teaching at Duke, he received Hicks's letter of November 2, 1948:

> Dear McKenzie, You'll probably have heard by now that your examiners did not make the recommendation we might have desired, but instead recommended that you should have the choice of a) resubmitting the thesis, or b), accepting a B. Litt. This is disappointing, but I hope you'll realize that the option of resubmitting is rarely given, and clearly means that the examiners did recognize the quality of your work. I'm sure of what the trouble was; as I always suspected myself, you did not have time to get that intricate stuff in the middle really lucid, and until that has been done the thing has an unfinished air. . . . My own feeling is that your best plan is to set to work to prepare your thesis for publication. If, a year from now, it seems to be coming into shape (considered from that point of view) it would be very good if you could put it in again, though of course that would mean coming over again to be re-examined and I appreciate that this would be a heavy burden upon you. If the thing does not seem to be coming straight in time, or if it is hopeless for you to be able to come over again, then I should take the B. Litt. I feel myself very strongly that your work has the real quality; but, as so often happens with Oxford doctorates, the two years is just not sufficient time to get the thing to the degree of finish which is required. (LWMP, 32, Graduate School)

After McKenzie wrote to the examination board about the requirement that he reappear for the viva (oral reexamination) and the near impossibility of his doing so, he received a letter from the University Registry:

> Dear McKenzie, I have consulted the social studies board about your difficulty in coming back for viva should you desire to have another try at the D. Phil. I'm afraid that the board was unwilling to ask for a decree exempting you from the viva voce examination on a second try. I am very sorry about this but I'm afraid it cannot be helped. This being so I now assume that you wish your original decision to stand and that you will take the B. Litt. I have told the college. I hope you are enjoying your job at Duke University and that you will not let this setback here unduly upset you, though I suppose it means that you'll unfortunately have to begin to work for a Ph.D. at home. (LWMP, 32, Graduate School)

The news was crushing. He was a married father, who believed that he was about to take a pay cut at Duke,[13] with no real prospects away from Duke except those for which an M.A. was sufficient qualification. A former valedictorian and Rhodes Scholar, McKenzie had failed academically for the first time in his life—even if D.Phil. failure at Oxford was not uncommon (Kenneth Boulding and Robert Clower are two other examples). McKenzie had already, prior to the Duke offer and its acceptance, "rejected an offer from Princeton of an instructorship, which had been arranged, I believe, by Frederick Lutz. I did this partly because I expected that Princeton would want me to pursue the Oxford thesis, which I did not wish to do. I think the event has proved my decision to have been correct" (McKenzie 1999, 2–3).

The Oxford thesis, a copy of which can be found in McKenzie's papers, is a curious piece of work. It is a nontechnical survey and "critique" of various theories in welfare economics. McKenzie attempted to provide his own theory, but it remains vague in its details: it has a very Oxbridge tone of disbelieving all work by others but putting

---

13 In fact, his pay was not cut. Instead, he was "frozen" at $4,000 for three years until 1951–52, when he was paid $4,500 (CBHP, 46).

forward little new itself. Two years later, in a letter to his Oxford friend Ian Little, McKenzie summed up the project saying:

> My [Oxford] thesis . . . was an attempt to get at the meaning of welfare theory. My procedure, after criticizing the utilitarians, was to develop all the assumptions which the new type of welfare economics requires if it is to make sense. This meant that I seemed to be developing a theory myself rather than criticizing the theories of others, and at the same time the necessary assumptions were so fanciful that the theory seemed of doubtful value. (LWMP, 32, Graduate School)

Away from Oxford, it did not read well, nor apparently did McKenzie ever want to see it again. He summarized the entire episode to his Keio audience in only two sentences: "I returned to my undergraduate university Duke as an assistant professor. This is a move much more typical of Japan than of the United States" (McKenzie 1999, 3).

## From Duke to Chicago to Duke

Returning to one's undergraduate college to teach, without a Ph.D., was not the best way to advance a scientific career in economic theory. While this course might have made good sense in England, where a solid "upper second class" honors degree from Cambridge could secure lifetime employment at a provincial university as late as the 1970s, and while it might have been appropriate for teaching in a denominational liberal arts college in the United States through the early 1960s, it was increasingly uncommon in American universities after the war. The reasons for this can be found on both sides of the academic job market for economists. For the sellers, the young economists, there was a generational bifurcation. Those trained before the war were unlikely to have had much technical—mathematical or statistical or econometric—skill.[14] Those who had not yet completed their training before the war, like Arrow and McKenzie, could

14 Sidney Weintraub, trained at NYU and LSE before 1939, had taken only one semester of college calculus before he was to teach economic theory at St. Johns University in 1947 and mathematical economics at the University of Pennsylvania in 1950 (Weintraub 2002, 208–45).

see that a new kind of economics was emerging during the postwar years. Perhaps, despite their own earlier marginalization, their future in economics might be different. That hope, never articulated until the later 1940s, was that the change in rhetorical conventions (what counted as convincing arguments to other economists) that the war effort had wrought might move economics to become a different kind of scientific discipline more closely connected to mathematics, applied mathematics, and statistics.

It was this academic world that both Arrow and McKenzie entered in the late 1940s. Stanford was to become a major university in that postwar period as California became the center of the aerospace and then later the computer industry. Duke, hindered by its location in a segregated South uncomfortable with Jews and Catholics, radicals and labor unions, and African Americans, was resistant to change and thus at a competitive disadvantage in the national job market. While one can readily understand the hiring of Arrow at Stanford and McKenzie at Duke, Stanford would not have hired Ph.D.-less McKenzie, and Duke would not have hired Jewish Arrow. In Duke's department of economics McKenzie would find only two men (there were no women in the department) of particular distinction, Joseph J. Spengler and Calvin B. Hoover, each of whom would eventually become president of the American Economic Association (AEA). Neither of them, nor any of their colleagues, had mathematical or econometric skills: Spengler was an important figure among economic demographers, while Hoover created the field of comparative economics and consulted continuously for agencies like the OSS and CIA. McKenzie had graduated only eight years earlier from Duke when he took up his appointment as an assistant professor, and consequently he had to work hard to reconstruct relationships with his former teachers, now his colleagues.

His main scholarly task in those first years was to return to his non-thesis Oxford readings, and tutorial papers, and rework them into at least one publishable article. Over the next two years he was able to write and then publish a paper in the *Economic Journal* titled "Ideal Output and the Interdependence of Firms" (1951). This Hicks-style piece, published not in *Econometrica* but in the house organ of the Royal Economic Society, established McKenzie as a bona fide economist even though the article eschewed the kinds of mathematical theory that were becoming more widely employed in the immediate postwar years.[15]

How, then, was McKenzie to develop his talents and nurture an ambition that had included winning a Nobel Prize in Physics? It was time to go north.

> [A]t Duke . . . I noticed a report from a meeting of the Econometric Society of a paper by Tjalling Koopmans describing his activity analysis. His work struck me as just the kind of theory I could have used in my paper on ideal output. This led me to apply to Jacob Marschak to visit the Cowles Commission in Chicago. To make this visit possible I received support from the Carnegie Foundation which devoted some resources to promote advanced education in the American South and also a fellowship from the Department of Economics in Chicago. (McKenzie 1999, 4)[16]

The Koopmans paper that interested McKenzie was developed for a transformational intellectual event for economic theory in the 1940s, the Cowles conference on activity analysis of June 1949, which we discuss more fully in chapter 5. The papers given there would eventually appear in a book edited by the conference organizer, Cowles research director Tjalling Koopmans (1951). It was with both excitement and a bit of nervousness that McKenzie sought a new direction. We can see tracks of both emotions in his handwritten letter to Ian Little from spring or summer 1950:

---

15 In the second paragraph of this paper, McKenzie states that he intends to show that a particular argument of A. M. Henderson will be shown to be "erroneous." Henderson of course was the outside examiner on his Oxford D.Phil. Revenge perhaps?

16 On August 29, 1950, McKenzie received a letter from the Rockefeller Foundation's General Education Board advising him that "the General Education Board has awarded you a fellowship in the field of economics for study at the University of Chicago. This fellowship provides for a stipend of $60 a month for not more than 12 months beginning approx September 15, 1950, and for tuition if needed in excess of your GI allowance" (LWMP, 6, Letters from Duke). Attached to the Rockefeller letter was the receipt from the University of Chicago bookstore for the purchase of the books by Birkhoff and Mac Lane (*Modern Algebra*), Feller, volume 1 (*Probability Theory*), and Courant, volumes 1 and 2 (*Differential and Integral Calculus*), for which he paid a total of $24.50.

> I abandoned the Ph.D. at Princeton, after Graham's death[17]... encountering Morgenstern's hostility[18], and getting saddled with Viner as a supervisor. Whew! My position here [at Duke] is not thereby jeopardized [by not having the Ph.D.] it seems. The purpose of the trip to Chicago is to get some fresh air, intellectually speaking, learn a bit of statistics, and perhaps make some kind of new start. I don't know. (LWMP, 6, Letters from Duke)

In 1950 McKenzie was thirty-one years old. Married, now with two children, he was trying to overcome his Oxford disaster and take his adult place among his former mentors at Duke. He was again seeking an entrée into a larger intellectual life.

It is clear that McKenzie had the support of his Duke colleagues for this move. Hoover, seeking fellowship support for McKenzie's year, wrote to the director of the General Education Board that if McKenzie could study econometrics at Chicago he might be able to teach this material at Duke: "It would be a good investment all around. Spengler tells me he believes McKenzie has one of the best minds of anyone whom he knows. I am inclined to agree with his judgment" (CBHP, 46, McKenzie). Indeed, a couple of months later and before McKenzie would depart for Chicago, he appeared to be very engaged in matters of curriculum and pedagogy. Appraising his teaching career so far at Duke, on June 3, 1950, he sent Hoover the remarkable document "Proposal for Reform of the Economics Curriculum," which set out his own vision of what an economics undergraduate education might consist (ibid.). Its features would eventually lead him, seven years later, to seek to implement that vision at Rochester. Because the "Proposal" gives voice to his thinking, it

---

17 On September 24, 1949, Graham fell to his death from the stands "at Palmer Stadium at the close of the Princeton-Lafayette football game" (Lester 1950). This account is in contrast with the remarks about Graham in Charles Kindleberger's (1991) autobiography. There, in a paragraph that begins "I have since wondered whether Professor Graham was in the incipient stages of a breakdown," he concludes with "Graham later committed suicide by jumping off the upper deck of Palmer stadium in Princeton" (129).

18 There is no other mention of this "hostility." One reasonable interpretation of the word is that Morgenstern, then as later, was impatient with neoclassical theory and wanted his students to do empirical work.

is worth considering it if only in outline since we do not have so many other clues to McKenzie's beliefs in the period before his year at Cowles.

McKenzie proposed a separation of business and economics classes, including a top-down approach to the latter as is common today. Recognizing that Duke's department and major was one of "economics and business administration," McKenzie suggested that its business-interested students be driven "either into business administration or out of economics entirely" since "the method of instruction is one that goes a very little way towards developing either the interest or the potentialities of the better students" (CBHP, 46, McKenzie). He proposed that the results of the first exam of the "Principles" course be used to locate the stronger students, who would then be excluded from class meetings to instead work closely, in seminar-tutorial fashion, with the instructor who would be able

> to lead them far more deeply into the subject. . . . Time which is devoted to the more capable students must be taken away from the less capable students. My second proposal is that [the course in economic theory] be made into a two semester course and be made a required subject for all majors in economics. . . . [It] would give some guarantee that those who took degrees in economics actually had some knowledge of economics. (ibid.)

McKenzie was clearly frustrated by average business-interested students taking an economics major consisting of an introductory course and a nonhierarchical collection of elective courses ranging from the "history of labor unions" to the "economic geography of Africa" to "economic statistics" (Craufurd Goodwin, personal communication). He believed that there was a central corpus of economic analysis the mastery of which defined an individual as an economist. He believed that educating a person as an economist required instructing students in such tools and techniques and modes of thought that could then be "applied" to the study of labor, or business cycles, or demography. Today economics at Duke, and at most other places in the United States and around the world, is taught in exactly that fashion: students are first taught the "core" of microeconomic theory, macroeconomic theory, and econometrics before they are permitted to take elective subjects. McKenzie's vision was clear and forward looking. That of his colleagues was not. His proposal was considered and put aside. It would be two decades before it would be implemented, more than a decade after he had left Duke. At Duke, McKenzie was a very lonely, very frustrated "Young Turk."[19]

This frustration contrasts sharply with the lofty tone that McKenzie adopted when speaking about the time he spent at Cowles. That period constitutes an actual, and very positive, "reminiscence bump" (see Mackavey, Malley, and Stewart 1991; Weintraub 2005). He was taught by the most sophisticated mathematicians of the day (Irving Kaplansky, Paul Halmos, and Saunders Mac Lane), and he even made a first, if indirect, impact on the shining star of mathematical research at the time: John von Neumann. In his Keio narrative he wrote:

> After two years teaching at Duke, in the fall of 1950, I worked at the University of Chicago as a graduate student in economics. The time I spent at the Cowles Commission was decisive in setting the character of my research career. At least half of my time, which comprised four quarters or twelve months was devoted to mathematics. I had three marvelous young teachers there, not very different in age from myself. These were Irving Kaplansky in algebra, Paul Halmos in measure theory, and Saunders Mac Lane in topology. I also benefited from a course in mathematical statistics with Jimmy Savage. Savage was writing his book on personal probability and I was able, along with Jacob Marschak, to call his attention to the work of Frank Ramsey which I'd encountered in England. Incidentally, I was indirectly responsible for calling this work to the attention of von Neumann and Morgenstern by way of my friend Ansley

---

19 This particular frustration did not abate after he returned from Cowles. To give a sense of how McKenzie viewed the educational practices of his colleagues, he saved a carbon copy of a note he passed out to all members of the economics department present at a faculty meeting in May 1952: "Gentlemen: I am not entirely satisfied with my remarks in this meeting. I should like to add the following assertion: '*As long as we have the textbook, classroom, compulsory attendance, five courses per semester* system of instruction, we are going to have bored, disinterested [*sic*] students. And, by jove, I don't blame them. This will be true for all departments and all the years.' L.W. McKenzie" (LWMP, 18, Currriculum).

> Coale. Ansley reported that von Neumann expressed no surprise that the work existed but wondered why he had not succeeded in finding it. Morgenstern on the other hand questioned whether Ramsey had used an axiom system. Of course he had. (McKenzie 1999, 4)

McKenzie had made it. He was in. He had a place in the community that would appreciate the kind of economic theory that he wanted to do. In the classes he took from Tjalling Koopmans, research director at Cowles at the time, and Jacob Marschak, the preceding director, he would meet the people with whom he was to be compared for the rest of his life.

> In economics I attended the classes of Koopmans on activity analysis and econometrics and of Jacob Marschak on decision making under uncertainty. My companions in these courses included John Chipman, Martin Beckman, and Edmond Malinvaud. Gérard Debreu, Karl Brunner, and Harry Markowitz were also in the Cowles Commission group, as well as Leo Hurwicz for awhile. I attended no classes given by a regular member of the economics department. (McKenzie 1999, 4)

McKenzie had begun his new career at the place where Arrow had become known to the community of mathematically and statistically sophisticated economists, and where Debreu was to find a community of like-minded scholars. In part 2 we will examine the nature of that community in more detail. But first we introduce our third protagonist, Gérard Debreu, whose own path to Koopmans's class was likewise strewn with obstacles.

# CHAPTER 3
## DEBREU'S SILENCE

Gérard Debreu was, in many respects, the odd man out. It was not only that he was French and not American but also that both his family and his education differed significantly from Arrow's and McKenzie's. In order to understand his role in the existence proof story, personality matters. Unlike both Arrow and McKenzie, Debreu was introverted, silent, shy, self-protective, and rather dogmatic. During research seminars, Debreu hardly ever spoke to pose a question—and when he did, he already knew the precise answer. He never debated with others; he only informed others about his final and incontestable thoughts. In 1972, at the AEA lunch in honor of Arrow's Nobel Prize, Debreu gave a toast. He noted "the breadth of your interests and your extraordinary willingness to discuss economic ideas at any stage of their development" (GDP 14, Arrow). In both such respects, Debreu could not have been more different from Arrow. Steve Goldman, a Ph.D. student of Arrow's and longtime colleague of Debreu's at Berkeley, speculated about how they communicated:

> I have visions of conversations consisting of Arrow going on at an extremely rapid pace, and Gérard every once in a while saying something, or coming back with a proof. I can't imagine a dialogue between the two of them. Their styles were just so different. Arrow was going to conjecture about anything. . . . He was unbelievably fast, but open in what his thinking was. Gérard's thinking was never open. (Steve Goldman, personal communication)

Debreu was extremely private even within his family. His lifelong friend Werner Hildenbrand said, "He could not show emotions. That was something terrible for him."

> Françoise Debreu: There was nobody really close to Gérard. You know that my husband was a very difficult man to get through to. He was not talking very much. He was thinking things inside and not exteriorizing at all. And for long time that was the main thing that frightened me.
>
> Chantal Debreu: My father also had a habit of staring at you. He would just sit and stare and never break eye contact. It was spooky. . . . He could be quiet for hours. Over a dinner, it wasn't unusual for him not to say one word during a three-hour dinner. (Françoise Debreu and Chantal Debreu, personal communication).

Debreu's silence created a distance difficult to bear for those who felt close to him. "If I tried to talk to him, simply chatting—that was impossible," Hildenbrand recalled. "When we had a walk, it could happen that twenty minutes might pass without a single word. I almost got mad about it. For me, with every step, it got more and more difficult to bear" (Werner Hildenbrand, personal communication).

His distant manner also shaped his academic career. His commitment to eschew the personal, to remain anonymous, was consistent with the emergent postwar academic culture: reticence could be associated with scientific detachment and professional integrity, and secretiveness could be an epistemic virtue, not a personal vice. It was not that Debreu was hiding personal or political biases by his silence; it was simply that silence was his way: "In a profession that rewards brash scholars who overstate the importance of their results, he [Debreu] rose to the very top by making no claims for his work beyond the unvarnished statements of his theorems" (Anderson 2005). His quietness thus expressed an authority that made Debreu an intellectual, though not personal, father figure to a generation of mathematical economists. When his community celebrated his sixty-eighth birthday, Debreu's perfectionism was the centerpiece of Dieter Sondermann's admiring toast:

> Gérard, we have learned from you to prove our claims. We did not dare to confront you with vague ideas. . . . [Y]ou have set standards by your mere presence. And these standards were: rationality, rigor and beauty. Gérard has always been to us a model of perfection, who not only talked about rationality but also applied it to perfect his everyday life. Such . . . perfection creates a certain distance. I think all of us in the beginning had to overcome a certain shyness, which was connected with our admiration. . . . Let's raise our glass and join in singing "Happy birthday to you, dear Gérard." (GDP 5, Sondermann)

With respect to Debreu's role in the making of the existence proofs, his silence creates a quandary for the historian. It can be interpreted in two ways. On the one hand his reticence, his discreetness, and the limitation of his claims to those that could be supported with mathematical rigor were a manifestation of his self-protective nature. The anonymity of mathematics shields the mathematician's persona from being expressed in his work. As Debreu wrote in his last publication: "[E]ven though a mathematical economist may write a great deal, it usually remains impossible to make, from his works, a reliable conjecture about his personality, because, in particular, formalized expression has deprived an author of his literary style for several decades" (2001, 4). Mathematics hides the mathematician's personhood. We indeed suggest that Debreu's insistence on mathematical rigor was a way to earn personal recognition without exposing himself as a person.

The epistemic virtue of reticence, on the other hand, turns into its antithetical scholarly vice as soon as it silences the works of others: Debreu, as we are going to see, both silenced the differences between him and his collaborator Kenneth Arrow and silenced the similarities between his and McKenzie's work. He appeared to believe, as we will argue in chapter 6, that both his difference from Arrow and his similarity with McKenzie would have diminished the credit he had been accorded. Thus our reconstruction of Debreu's role will account for both the virtue and the vice.

## Mathematizing Oneself

Gérard Debreu was born in the English Channel port city of Calais in the summer of 1921. It is difficult to understand his intellectual development without understanding his early losses to which his career was a response. At an early age, a younger sister burned to death in her cradle in a room adjacent to his own. When Debreu was eight, his father committed suicide. One year later, his mother died. As an orphan, he was sent to boarding school where he had only rare visits with his maternal grandmother. Although he had several paternal uncles and aunts who could have taken him in, they did not because of the shame attached to children of a suicide. This shame weighed heavily throughout his entire life, so much so that he would never speak to anyone about these early losses, not even to his closest friends or family. His daughter found out about these events only after he passed away in 2004.

Having never acquired the language to reflect on his early losses, Debreu experienced his life as something he could never fully control, as something he did not actually conduct, but rather as something that happened to him—a "random walk" as he called his intellectual biography (1991)—in spite of the detailed five-year plans he constructed for himself. Debreu could not live up to the ideal integrity of a subject who conducts his life on equal footing with the world. In mathematics, however, this need for protection would be met. His daughter indeed associated his career ambitions with the early losses in the following way:

> He had this anxiety about control and lack of control, because he'd had a number of really horrendous events that happened to him in his life. . . . I think that explains why he was so socially inept. Where he tried to get his survival was by shining academically, which was something he could do. So he was always looking for where he can exercise control. (Chantal Debreu, personal communication)

On two occasions, when Debreu offered his condolences to friends after their fathers' deaths, he also referred to their careers: "I hope that your sorrow is made a little less heavy," he wrote to Graciela Chichilnisky on March 3, 1971, "by the thought that you are living up to what must have been some of your father's ideals" (Debreu Papers, additional carton 3). He knew neither her father nor his ideals. To Herbert Scarf, he wrote on December 7, 1964, "I want to offer you all my friendship in your sadness. Your career must have been a great reward to [your father]" (Debreu Papers, additional carton 1).

Debreu's career began at secondary school, the "college" in Calais. Teachers were authoritarian and learning was a competition: once a week, for example, pupils had to change seats according to their test grades. He was an eager student, interested in the sciences, in particular physics, and soon got to know the "austere beauty of mathematics" (Debreu 1991, 3). A major event of his youth was the *concours général* in which he won the second prize in physics, a boat trip to French West Africa. To be positively selected is to be apart, not as one shunned but as one appreciated, honored, and celebrated. Having recognition from others mitigates somewhat one's experienced alienation. Such a positive feeling hints that similar good feelings could follow on similar future successes.

In 1939 he took the Baccalauréat and left Calais to enter a preparatory class for the Grande École entrance examination. In normal times that class would have been held in Paris, but the German occupation interfered. Instead of going to Paris, he was sent to study (using an improvised curriculum) in Ambert and Grenoble in the Free Zone administered by the Vichy government, where children were little safer from the tumult of the north. "The isolation of the Ambert novitiate," he recalled, "often made it possible to forget that France was at war" (Debreu 1991, 3). Debreu remained both ambitious and successful. After his two years of preparation for the entrance examinations, in 1941 he won the competition for admission to the École Normale Supérieure in Paris on rue d'Ulm.[1] For those aspiring to a career as a scholar, the École Normale Supérieure was

1 For those unfamiliar with the French system, a Baccalauréat permitted a French student to enroll in a French university. These were generally very large institutions. However, there was an elite set of institutions like the École Polytechnique, the École des Mines, and so on that prepared a student for the highest civil service and government positions.

the Promised Land. In Debreu's time, admission was offered to fifty students in each of the humanities and the sciences. Of the science students, around twenty were in mathematics.

Life in rue d'Ulm was obviously affected by the German forces occupying Paris. In 1943 Debreu had to do forced labor as a *terrassier*, rebuilding streets and bridges for the German troops. The image of Debreu's body mucking about in the raw earth, surrounded by coworkers from all classes, all under the direction of the Nazis, was always startling for his daughter. Stiff and austere as her father was, she as a child imagined he slept "fully clothed in his dress shirt and bow tie" (Chantal Debreu 2005). Debreu took big risks when escaping from this work in order to attend classes at the École Normale, which, until D-Day, were never interrupted. Attending classes provided some protection against the rest of occupied Paris. It represented another world, as he later described in a speech at that school after receiving the Nobel Prize (June 1, 1984):

> Paris from 1941 to 1944 was darkened in many ways. The blackout, the curfew, nights interrupted by sirens, BBC news distorted by radio interference, the difficulty of making one's way around the city, everything contributed to creating on rue d'Ulm an image of permanent implosion [*une pression qui paraissait le maintenir en état d'implosion permanente*]. An intellectual tension reigned in this universe, which I have never experienced elsewhere. (GDP, 14, École Normale, translated by the authors)

This tense pressure on rue d'Ulm intensified and accentuated the feeling of separateness and specialness that permeates the fabric of such elite institutions. In this "superheated intellectual atmosphere" (ibid.), coupled with a "vivid feeling of solidarity" (ibid.), Debreu came alive intellectually. He found a source of satisfaction that placed him far beyond the surrounding world and the events that had previously defined his life.

> Entering the École Normale Supérieure in the fall of 1941 meant another initiation. The three years during which I studied and lived at the École Normale were rich in revelations.

> Nicolas Bourbaki was beginning to publish his *Eléments de Mathématique*, and his grandiose plan to reconstruct the entire edifice of mathematics commanded instant and total adhesion. Henri Cartan, who represented him at the École Normale, influenced me as no other faculty member did. . . . The new levels of abstraction and of purity to which the work of Bourbaki was raising mathematics had won a respect that was not to be withdrawn. (Debreu 1991, 3–4)

Debreu's teacher Henri Cartan, the oldest founding member of the Bourbaki collective, shaped his image of mathematics. He also shaped his intellectual ethos insofar as Bourbaki mathematics "commanded instant and total adhesion" (Debreu 1991, 3).

The Bourbaki program and its role in the history of mathematics has been explored by Leo Corry (1992), whose work has informed the account of Debreu's Bourbakism in economics by Mirowski and Weintraub (1994) and Weintraub (2002). For the present narrative, we can ignore the larger history of the workings of Bourbaki to focus on the intellectual attitude that Debreu inherited from Bourbaki, a particular mathematical sensibility that made Debreu different from both Arrow and McKenzie. During Debreu's Nobel lecture, Debreu indeed explained the difference between him and Arrow, by referring to Bourbaki:

> Kenneth Arrow has told in his Nobel lecture about the path that he followed to the point where it joined mine. The route that led me to our collaboration was somewhat different. After having been influenced at the École Normale Supérieure in the early forties by the axiomatic approach of N. Bourbaki, I became interested in economics toward the end of World War II. (Debreu 1984a, 268)

What made the young Debreu so committed to the mathematics of this collective? As discussed elsewhere (Weintraub 2002), the new vitality of French mathematics was evident by the mid-1930s. With the virtual elimination of an entire generation of French mathematicians in World War I's carnage, those coming into the field in the 1920s had no mentors other than those trained in the nineteenth century. As a consequence several of those young mathematicians, connected through their training at the École Normale Supérieure but having first jobs teaching in the provinces while living in Paris, began a project to reconstruct the mathematics curriculum, and their own out-of-date textbooks, by writing a new mathematics from the beginning. They called themselves "Nicholas Bourbaki" after an obscure French general from the nineteenth century who had been parodied in a prank at the École Normale Supérieure in their student days.[2] Their rules for Bourbaki were simple: all decisions had to be unanimous and made at their meetings one to three times a year in what they called "congresses," new members had to be vetted in a trial period, members had to resign at age fifty, and everything about Bourbaki had to remain a secret. Their book project, which came to be called *The Elements of Mathematics*, attempted to build mathematics up logically as an edifice of ever more complicated "structures": topological, order, and algebraic. This project was inflected with a perspective from their own training at the École Normale Supérieure that had eschewed applications of mathematics and its connections with the natural sciences—the mathematics students at École Normale studied mathematics, not science. As a consequence this was a new mathematics of the purest sort, one quite separate from any of the physical problems that had given rise to the "structures" in the first place (see Corry 1992).

The Bourbaki collective's insistence on the anonymity of mathematics must have been liberating for each of the mathematicians. Rather than each single member speaking out in his own name, they aimed at letting *mathematics speak for itself*. The secrecy concerning the list of members allowed Bourbaki to become a symbol not of a particular school of mathematics but of mathematical rigor itself. Their world of mathematics was incontestable and sublime. By hiding the social process of its making, Bourbaki mathematics was self-speaking, self-contained, and invulnerable (see Livingston 1999; Gray 2008). Debreu likewise never promoted a particular school of mathematics. He was never an outspoken Bourbakist in economics.

2 A good introduction to Bourbaki, and its history and contributions, is available in Mashaal 2006; see also Aubin 1997.

In his *Theory of Value* he would only speak of "the formalist school of mathematics" that informs the standards of rigor (Debreu 1959, 1). It is thus not surprising that the name of Bourbaki did not enter economists' new consciousness. Economists, just like Debreu, hardly ever discuss the reasons why they prefer one kind of mathematics over another.

But what made the anonymity of mathematics intellectually so appealing for Debreu? In the moment that structure and meaning are separated, and structures become the only object of concern—which is the imperative of the Bourbaki program—meaning is set free from humans' urge that it be articulated, exposed, or defended. Anonymity liberates intellectual life from its ground in messy human nature as might be experienced in discursive activities such as debates. It is less the attraction of disclosing and getting closer to something but rather more the liberation of aloofness from the mundane that nourishes the affects of a Bourbakian intellectual life. The axiomatic separation of structure from meaning discloses a new kind of joy—an "unsurpassed, addictive intellectual pleasure," as Debreu would say later (1984b)—beyond any epistemic trade-offs between the general and the particular, beyond compromises between induction and deduction, beyond the ups and downs of appearance and concealment that usually describe the search for knowledge. This intellectual experience produces an intensity of experience that many mathematicians believe transcends any other deployment of the human mind. It is this elevation that would lead others to speak of Debreu, as he was called in Frank Hahn's family, as "God."

A striking feature of the Bourbakian approach, and one that engaged Debreu, was its lack of interest in either scientific truth or philosophical justification. Mathematics, science, and philosophy are, as is required by their axiomatic methods, separate. The concern for truth and the concern for rigor exclude each other. Bourbaki mathematics was not designed for science but for mathematics itself. It was supposed to provide an axiomatic "foundation for the whole of modern mathematics" (Bourbaki [1939] 1968, v). Its purpose was the mathematization of mathematics and thus its liberation from a pragmatic context in science. Applying Bourbaki is, in other words, an oxymoron. For the same reason, the Bourbakists had to remain silent about a philosophical justification or explanation of their work on nonmathematical terms. Bourbakian Jean Dieudonné, when asked whether Bourbaki's "structures" might correlate with something "real," said:

> On foundations we believe in the reality of mathematics, but of course when philosophers attack us with their paradoxes we rush to hide behind formalism: "Mathematics is just a combination of meaningless symbols", and then we bring out Chapters 1 and 2 on set theory. (Dieudonné 1970, 145)

There cannot be any such thing as an explicit philosophy of Bourbakian mathematics or Bourbaki would have had to tackle "the philosopher's paradoxes" and take a position about the philosophy of mathematics. This positioning, though, would have to take place outside the frame of "Chapters 1 and 2 on set theory" or outside the "universe of discourse . . . explicitly listed at the outset," as Debreu requires at the beginning of his *Theory of Value* (1959, 3). The problem of Debreu's Bourbakism, accordingly, was not a particular philosophical belief about the role of mathematics in economics or any specific belief about economic reality. Debreu never learned to enjoy discursive, explicative, let alone contestable intellectual activities. As we will see, this estrangement from the give-and-take of normal intellectual conversation would complicate his engagement in a joint project like his collaboration with Arrow.

If the intrinsic joy of practicing mathematics becomes the main engine of research, note that this joy comes at the cost of frustrating the need for living in a comprehensible world, a need that commonly motivates intellectual practices. Pragmatic standards such as social significance as well as philosophical standards such as empirical truth lose relevance. The research payoff, apart from intrinsic joy, is thus not the concrete relevance of one's work but must be itself formal: temporal priority. If the only objects of mathematical practice are "structures" themselves, not their meaning, so too the reward of mathematical practice can only be framed in terms *other* than its meaning. Issues of priority, even if they have always existed in mathematics, are therefore reinforced by the intellectual purism of

Bourbakism. Indeed it was Debreu who was more concerned about priority than either of the two economists involved in the construction of the existence proofs.

## Half-heartedness

The end of the war coincided with the end of Debreu's life as a student. The exigencies of the approaching postwar period for France, and his own leave-taking from the rue d'Ulm enclave to seek employment, left him to struggle with his faith in the "grandiose edifice" of Bourbaki. The end of the war stimulated a new interest in the social sciences, in particular in the question of what holds society together in the absence of a leader. This led Debreu away from mathematics. But how was he to engage with such an interest?

The reasons why Debreu left mathematics remained obscure to him. When he was later asked why he left, his immediate reply was "the reasons aren't clear to me" (1987). Nor did Marcel Boiteux, who would later become a colleague of Debreu's at the Centre National de la Recherche Scientifique (CNRS), ever understand why it happened (personal communication). Here are the telling lines in which Debreu described, in hindsight, how he entered economics:

> [B]y the end of 1942, I began to question whether I was ready for a total commitment to an activity so detached from the real world, and during the following year I explored several alternatives. Economics was one of them [he considered astrophysics, too, but his teacher, who was Jewish, had had to flee France]. In 1943–44 the teaching of the subject in French universities paid little attention to theory, and the first textbook that I undertook to read reflected this neglect. The distance between the pedestrian approach I was invited to follow, and the ever-higher flight I had been riding for several years looked immense, perhaps irreducible. Reason counseled retreat to a safe source. What kept me on an unreasonable heading? The formless feeling that the intellectual gap could be bridged? The wishful thought that the end of the war was near, and the perception that economists had a contribution to make to the task of reconstruction

> that would follow? An improbable event brought my search to a close. Maurice Allais, whose *À la Recherché d'une Discipline Economique* had appeared in 1943, sent copies of his book to several class presidents at the École Normale. (Debreu 1991, 3–4)

Debreu could have easily rationalized his decision to enter economics by the historical situation. But he did not do that. He made the decision in light of, but also against, his concerns for the social world around him. He felt "detached from the real world" because of the pressing political situation in which the Bourbakist project seemed inappropriate. But given his intellectual values, he could experience such impulse merely as a "formless feeling," a "perception" he could not take seriously. He felt the impulse to move on but was repelled by the thought of so doing. He had not learned from the Bourbakists how to reflect on intellectual engagements with a "real world." He had learned rather to be suspicious of such engagements that seemed to produce an "unreasonable heading" of "wishful thought."

Did he solve the dilemma between his intellectual values and his social concerns? As his reflection suggests, he did not. Instead he entered economics by a chance encounter with an authority: Maurice Allais, whose book, though hardly a model of Bourbakian rigor, led him to believe that his mathematical training could be of use after all. Following an authority, Debreu did not decide to enter economics; he simply had not acquired the intellectual wherewithal to make such a decision for himself. He entered economics half-heartedly.

The path, though, was not yet set. In the spring of 1944 Debreu was supposed to take the *agrégation de mathématique* in order to qualify for an academic position as a mathematics teacher in the French system, but D-Day disrupted those plans. He had to go into the French army, first at officer school in Algeria and, after May 1945, for six weeks in Germany, his only "opportunity to experience a life outside the academic cocoon" (Debreu 1991, 4). At the end of the war, he faced another weighty decision: he proposed marriage to Françoise Bled, whom he had been dating for three years. But in this case, too, it wasn't really his own decision: "When he proposed to me, it was not a surprise. He was pushed a little bit by some of his friends. . . . Finally he did it, a little awkwardly" (Françoise Debreu, personal communication). After the wedding, Debreu began studying for his *agrégation* to be taken at the end of 1945. In August 1946 their first daughter, Chantal, was born.

With a degree from the École Normale he easily got a job from Allais as an *attaché de recherches* at CNRS. While there he taught a course on business cycles at the École d'Application de l'Institut Nationale de la Statistique et des Etudes Economiques, and from Allais he learned of the Walrasian system. But he could neither live out his mathematical taste that Allais, the engineer, did not support nor hope for an academic career, since Allais was not supported by the French economics community: mathematics was viewed with suspicion by French economists. Allais, located at the École Polytechnique and known as a "mathematically mad person," was far from being a celebrity in French academia (Debreu, in Weintraub 2002, 137). He could not provide much guidance. According to Boiteux, who spent these three years at CNRS working next to Debreu, their main joint activity was less the discussion of their work but engaging in weekly physical exercise in Parc de Saint-Cloud (personal communication). Left alone with the economics of engineers, Debreu had no idea how to continue his career. Many years later he recalled that he had the option of going into industry since others around Allais, the so-called French engineers, came to be attached to Électricité de France: "It was more or less clear that I would be an economist, and that I would not be a professor of economics at a university. Possibly I had in mind vaguely jobs like, let us say, Boiteux's job" (Debreu, in Weintraub 2002, 139).

The work of the French engineers indeed formed Debreu's implicit image regarding the future of economic theory. Edmond Malinvaud, who would be friends with Debreu for the rest of their lives, remembers a lunchtime group where they read, for example, Abba Lerner's *Economics of Control* (see Krueger 2003, 184). There Debreu got to know some of the left politics of mathematical economics. It must have been from this experience that he spoke later of the French need for "reconstruction" to be made (Debreu 1991, 3–4), which the others of this group actually engaged.[3] In his first economic article in French (Debreu 1949)—still using calculus instead of convex analysis—he showed considerable respect for the interpretive dimensions of general equilibrium theory. Here, for the only time in his life, he made the reader aware of a "certain danger" regarding the welfare interpretation of a Pareto optimum that once was at the heart of the socialist calculation debate: he warned of the "risk of letting pass an absolute that is eminently relative" (ibid., 614, translated by the authors).[4]

Two and a half years after his *agrégation* Debreu had his first contact with American economists at the "Seminar in American Studies" in Salzburg. There he found new hope for an academic career. The seminar had been constructed to acquaint European economists with recent developments in economics in the United States and to reestablish links between economists in the postwar period. Debreu proposed a paper on Pareto optimality to Wassily Leontief, who co-organized the seminar. It was accepted. In it he expressed what he considered the standard political use of the Walrasian model. He wanted "to clear the aim and the means of economic planning" (Debreu to Leontief, July 5, 1948).[5] Once he arrived in Salzburg he learned of a new research environment that would support an economics with more advanced mathematics than was possible with Allais. This was less a result of meeting with Leontief and Solow, who participated in the seminar, but rather because he found out about the *Theory of Games and Economic Behavior* by John von Neumann and Oskar Morgenstern. That 1944 book made exciting use of axiomatic mathematics and convinced him to apply for a Rockefeller fellowship to study in the United States in 1949. He was duly accepted and spent most of his first months in the Harvard library continuing

---

3 This was the only group the interests of which Debreu would later support by his personal fame. In 1994, during the negotiations between EDF and Framatome, a company building nuclear plants, he sent letters to Jacques Chirac arguing for state control (GDP, 10, EDF). But Boiteux, president of EDF from 1967 to 1987, had no idea why he did so (personal communication).

4 Yet he never learned how to express his intellectual passion in differential calculus and continued work in the Bourbakian sphere. He translated de Finetti (1949) without publishing it (GDP, 9).

5 We thank Roger Backhouse for providing this letter to us. Wassily Leontief Papers, Harvard University, Manuscripts and research notes, 1930–70, Box 8, Folder: Salzburg Seminar: Applicants and Students.

his reading of the *Theory of Games*,[6] though he occasionally attended a lecture by Joseph Alois Schumpeter on economic theory. He spent the summer at Berkeley and visited the Cowles Commission in October: "During my visit to the Cowles Commission in the fall of 1949 I did not have close contacts with the faculty" (Debreu to Weintraub, December 7, 1981). Nevertheless, the talk he gave on that visit must have made a considerable impression on Koopmans since he was immediately offered a position as a research associate beginning in June 1950 (Debreu, in Weintraub 2002, 142).

## Discretion

The Debreu family arrived in Chicago in August 1950. They were warmly welcomed. When his wife, Françoise, took the boat from France to the United States with her two babies, onboard she was recognized by Eveline Weil, the wife of the Chicago mathematician André Weil, one of the leading Bourbakists (and brother of Simone Weil). Though she and Françoise had not been close in Paris, Eveline Weil had known of the Debreus coming to Chicago, and the two women quickly became good friends. Their husbands would become closer, too. André Weil, at Chicago since 1947, enjoyed having Debreu in Chicago as a university colleague with whom he could speak both French and mathematics. Debreu also developed cordial relations with other Bourbakists at Chicago, most notably Saunders Mac Lane and Armand Borel, as well as other young group members who frequently passed through the department in Chicago, such as Samuel Eilenberg and Serge Lang. It was through Weil that he also remained in contact with members back in France such as Pierre Samuel, Jean-Pierre Serre, Jean-Louis Koszul, and later Laurent Schwartz.

During the 1950s the Bourbaki group's influence in U.S. mathematics departments began to grow though, to be sure, mathematical purity had never been as prized in American academic culture as it was in France. Thus Debreu, in the Cowles community, was privileged

---

6 Boiteux applied, too. Both were selected, and Allais had to decide who would go to the United States first. Debreu always emphasized this decision as determined by chance. They tossed a coin: "La pièce tourna longtemps sur elle-même sur la table et finit par décider que je quitterai Paris a la fin de 1948" (GDP, 14, École Normale). He must have told this story on so many occasions that this was the first thing his wife told Düppe when he began talking to her.

to have special access to the leading Bourbakists. Consequently in economics Debreu uniquely represented the mission of the Bourbakian "working mathematician":

> At Chicago, there is an excellent mathematics department, very rigorous, which has been strongly influenced by the Bourbaki school, which was mainly represented by the great mathematician André Weil. . . . Personally, as soon as I arrived in Chicago, I felt this influence. (Debreu, in Bini and Bruni 1998, 19, translated by the authors)

Tjalling Koopmans, who had just become director of research, replacing Marschak, particularly welcomed Debreu's Bourbakism. Debreu's strengths aligned him with Koopmans's own efforts to redirect Cowles's focus toward economic theory: the shift was made manifest by the change in its motto from "Science Is Measurement" to "Theory and Measurement." For both changes, the replacement of "is" by "and," and of "science" by "theory," Debreu would be vital. Upon his arrival Debreu thus did not notice much of the former, empiricist spirit of the place. Klein had left years earlier, and the activity analysis conference that we discuss in detail in chapter 5 had been held in June 1949. Instead he felt liberated from the intellectual suspicion that surrounded would-be practitioners of mathematical economics in France: "Whereas before I was in a group which felt mathematics went too far and points of rigor were not terribly important, at Cowles I came to think, very quickly, that full understanding of a problem required no compromise whatsoever with rigor" (Debreu, in Weintraub 2002, 153).

But while Debreu's Bourbakism was welcome, he again stood apart. At Cowles, he found himself among many economists who were socialized to the scientific practices that came to be called operations research during World War II and who thus were conversant with the application of mathematics to military and economic planning. There, in a "very closely knit group" as he recalled, he met a number of economists who now had a contribution to make to the task of reconstruction he had perceived in postwar France (Debreu 1991, 4). But Debreu stood apart from those who would employ mathematics to remake the world. His intellectual socialization in mathematical purity was a protection from rather than a confrontation with the war and the postwar economic reconstruction. What he had learned from Allais was of little help in engaging with the cold war milieu of military strategy and planning for postnuclear exchanges: "I was left alone to do my work during the five years from 1950 to 1955, a marvelous opportunity that I tried to use fully" (Debreu, in Feiwel 1987, 256). For Debreu, the position at Cowles meant he had arrived in a place that was connected, as the next chapter will show, to the most central institutions of U.S. postwar science. Yet it meant at the same time a return to his Bourbakian cocoon.

# PART II
## CONTEXT

For surely, when we inquire into the context of . . . work like [Arrow's, Debreu's, and McKenzie's], we are interested above all in the extent to which that context provides resources for the productions of the kinds of meanings that the text displays to us. To have information about this aspect of the text's context would not illuminate the operations of [Arrow's, Debreu's, and McKenzie's papers] in their specificity, in their details as we follow or track the text's narrative. On the contrary, it is the other way around: the context is illuminated in its detailed operations by the moves made in their [existence papers].

**WITH APOLOGIES TO HAYDEN WHITE 1987, 212**

# CHAPTER 4

## SITES

The end of World War II reconfigured the institutions of U.S. science. While in continental Europe the conflagration and aftermath of the war froze optimism about reason and science, in the United States it created the mind-set that the scientific achievements that had enabled victory could similarly enable a prosperous peacetime society. The centrally planned wartime economies of the United Kingdom and United States had succeeded in defeating the Axis Powers. The war-ending shock of the August 1945 atomic bombs and the resulting Japanese surrender reinforced the idea that planned government support of science would sustain Western freedoms. Based on those wartime experiences, in the war's aftermath the Anglo-Saxon world had renewed hopes for the potential benefits for planning in a free society (Bockman 2011). The Atlee government's National Health Service and the Truman administration's Employment Act of 1946 reified the belief that planning—considered central governmental engagement with the market economy—could produce socially desirable outcomes.

American economists had worked in the planned war economy. In addition to the economists' usual jobs at the Departments of the Treasury, Agriculture, and Commerce, as well as the Office of Price Administration, economists worked on military applications of decision analysis. This new "operations research" ensured that interest in technical economics remained at a very practical level until the war's end. The operations research community, and the nascent game theory community, used newly developed tools in order to solve search problems (e.g., antisubmarine warfare), allocation problems (e.g., steel for production of tanks versus battleships), computational problems (e.g., code breaking), and bombing problems (e.g., low-altitude high risk-high gain versus high-altitude low risk-low gain). The programming analyses that emerged from operations research required well-formulated problems and computable solutions.[1]

Postwar understanding of the wartime role of scientific research reshaped universities' missions (Leggon 2001, 221–24). In 1944–45 the U.S. Congress commissioned a study of how scientific research should be funded in the postwar future. There appeared to be two possible models. The first was to have Congress set up a research agency and then fund or earmark projects prioritized by national needs, as had been done during the war by the Office of Science, Research, and Development (OSRD). The second was that Congress could give money to the scientists directly or through their employers, letting peer review and competition solve the allocation problem politically unencumbered by pork barrel politics. The former model meant continuity with the wartime regime; the latter meant a return to the ideal vision of the autonomy of science.[2]

The institutional design of postwar science thus required balancing maintenance of some elements of wartime scientific institutions and practices while simultaneously rejecting the elements that did not fit well with a democratic peacetime society. Universities were drawn into two not necessarily compatible roles: they were to host the "scientific community" that was supposed to exemplify the values of a free democratic society, but they also had to serve national purposes. The pursuit of truth among scientific peers was a model of the behavior that one would expect from a well-functioning democracy. This notion informed, for example, the report of Harvard's

1 Wartime applications of game theory and operations research have been well examined (in Mirowski 2002, chap. 4; Leonard 2010, chap. 12; Erickson et al. 2013; Erickson 2014).

2 One of the key science administrators, Warren Weaver, who was the head of the Applied Mathematics Panel of the National Defense Research Committee, argued in favor of maintaining continuity with the wartime organization of science: "The distinction between the military and the civilian in modern war is . . . a negligible distinction. . . . It may even be, for example, that the distinction between war and peace has gone by the board" (Weaver, in Collins and Kusch 1998, 253).

Committee on Higher Education in its report *General Education in a Free Society* (Buck et al. 1945), as well as Harvard president James B. Conant's *On Understanding Science* (1947). These writings, as Hollinger reported, "selected from the available inventory those images of science . . . serving to connect the adjective scientific with public rather than private knowledge, with open rather than closed discourses, with universal rather than local standards of warrant, with democratic rather than aristocratic models of authority" (1996, 444). But with the lifting of wartime secrecy, Americans learned that the success of scientists, and the technology they had brought to the war effort, had created a large government "owned" scientific community that had worked with a sense of national purpose: Oak Ridge, Hanford, and Los Alamos were the outward manifestations of this new scientific world. These institutions hosted science in ways quite apart from the democratic ideals of open universities—classified research remained taboo in universities. They, too, had to find support in postwar science programs.

The continued success of science in the creation of the postwar society would thus depend upon a resolution of the tension between continuity and reform, between transparency and secrecy, between scientific control of society and traditional values like liberty, between the autonomy of open science and its emergent social role as preserver of democracy against its enemies. The A-bomb, to put this tension bluntly, was a product that secured the victory of Western liberty but was at the same time a product of this very undemocratic secret scientific world. All academics in the United States after 1945, and thus our three protagonists, experienced to a greater or lesser degree some psychological tension as they internalized these conflicting values of science (Merton 1963, 276).

The institutional arrangements that emerged did not resolve this conflict. In the end, Vannevar Bush's report *Science, the Endless Frontier* (1945) presented Congress with a mixed model for support of postwar science. As a later historian noted,

> There is no consensus on the origin of this report . . . [but] there is consensus that the report was written in a highly charged political context—the growing debate over how the federal

> government "should advance science for the general welfare in peacetime" (Kevles 1996, 5).[3]. . . Significantly, this [Bush] report linked the fortunes of science to those of the government: "Since health, well-being, and security are proper concerns of government, scientific progress is, and must be, of vital interest to the government" (Bush 1945, 11). The report acknowledged and gave official sanction to the unchallenged right of the scientific community to manage its own affairs. This meant that the government agreed to let the scientific community decide how funds were allocated and the scientific assessment of their use. The definitive characteristic of this contract was that scientists controlled science—how it was done, by whom, and under what conditions. (Leggon 2001, 223–24)

Bush had been the director of the OSRD. That office was closed in December 1947. In its place Congress would fund a National Science Foundation with an annual budget appropriation, and the NSF would make grants to scientists through a peer review process organized by disciplinary scientific panels.[4] Since the federal government would remain outside this process, only universities had the organizational weight to administer the NSF grants. This model resulted in an increased competition between universities that had to hire individuals who were likely to attract these funds. And in the immediate postwar years, these were the same scientists who personified its success during the war and who had previously been paid by the OSRD.

At the same time government could not give up the military infrastructure it had built up during the war. National security appeared to require that the government continue its direct funding of specific

---

3 Historians of economics, and historians of science more generally, have only recently learned that Paul Samuelson claimed to be one of the three actual authors of the report (Samuelson 2009, 83).

4 In his recent study on the cold war history of social science patronage, Mark Solovey (2013) describes the debates that led to the creation of the NSF, notably the involvement of the institutionalist Wesley Mitchell and the agricultural economist Edwin Nourse as representatives of the Social Science Research Council. Mitchell urged, in contrast to Bush, the inclusion of the social sciences not only on unity-of-science grounds but on the basis of their ideological neutrality were they to be treated as real sciences. It should be clear that Mitchell's position in the NSF debate was not unrelated to his position in the so-called measurement without theory debate with Tjalling Koopmans at the same time.

research projects on, for instance, nuclear and thermonuclear weaponry. The Hiroshima and Nagasaki bombs had created a science race among nations. Congress thus sponsored national laboratories that engaged in largely secret defense-related work. In addition, Congress created administrative entities in support of national defense objectives. It authorized funding for the Office of Naval Research (ONR), the Atomic Energy Commission, the National Advisory Committee for Aeronautics (later NASA), the Defense Applied Research Projects Agency (DARPA), and the Air Force's Research and Development center in Santa Monica, an initially private corporation that would come to be known by its acronym, RAND.[5] These institutions produced classified research that shaped the cold war production of knowledge. The scientists hired by these institutions were the same individuals who, to a large extent, attracted the most NSF funds for their university departments. The two sites of science, inside academia and outside academia in national laboratories, were inhabited by roughly the same scholars. Kenneth Arrow was only one of them.

In the previous chapters our narratives reconstructed the beginnings of the scientific careers of our three protagonists, the acquisition of their scientific selves. In this chapter we will step back and look "outside" these individuals to the postwar sites where their work would find an audience. That audience would itself become a community in the sense of what Karin Knorr-Cetina called an "epistemic culture" (1991).[6] We speak of "sites" since this new audience had several institutional locations that were only loosely connected: the Cowles Commission, the RAND Corporation, the two departments of mathematics and economics at the University of Chicago and Princeton University, and the Institute for Advanced Study in Princeton. These sites had different values. In the circulation of individuals

---

5 "A song said to have been popular among scientists returning to their academic teaching and research responsibilities in 1946 had the title, 'Take Away Your Billion Dollars' " (Sapolsky 1990, 35). Sapolsky's superb volume details the navy's own support for science just prior to, and during, the war and shows how the ONR emerged from the mix of navy needs, the ambitions of young scientifically trained navy officers, the navy's own hierarchal command structures, interservice rivalries, and congressional compromises.

6 We are not here employing the full scope of Knorr-Cetina's idea; the same kind of notion could be associated with Ludwig Fleck's "thought collectives" ([1935] 1979) or Stanley Fish's "interpretative communities" (1980).

and ideas around these sites the existence proofs took on the symbolic meaning of heralding a new era in economic theory. In chapter 6 we will examine the event that brought together many of the individuals associated with these sites, the Cowles Commission's activity analysis conference of June 1949. With that event understood, we will be able to track the effect of the new ideas on the particular work that would bring credit to our protagonists.

## RAND

When OSRD shut down after the war, the military still had research interests and resources to pursue those interests. The RAND Corporation was founded in May 1948 as an independent think tank to carry out research for the military. It was located in Santa Monica, California, near the beach. Its mission was "to insure the continuance of teamwork among the military, other government agencies, industry, and the universities."[7] The research carried out at RAND was eclectic: it comprised weapons systems engineering, abstract mathematical inquiry, "basic" research in logic, computational technology, operations research, and game theory, much of this under the label of "systems analysis." RAND was a socially committed transdisciplinary environment without immediate connection to any of the traditional concerns of economics. In 1949, only 3 percent of those employed at RAND were economists (Mirowski 2002, 209). Kenneth Arrow shared the RAND vision of applying the principles of rational agency to politics and warfare. He believed that economic systems were expressions of the rationality of its agents. After 1947 RAND became his second intellectual home.

Some historians of economics have emphasized the role of RAND in the transformation of postwar economic science. In particular Mirowski (2002), Amadae (2003), and Leonard (2010) have brought forward comprehensive historical research and analyses of the connection between RAND and the emergent developments and applications of the theory of games to problems in national defense planning and strategy. These scholars draw different lessons from the history, as Mirowski and Amadae suggest that postwar economics was a cre-

7 http://www.rand.org/about/history.html.

ation of, and a combatant in, the cold war. It is our belief that RAND has been inappropriately demonized by historians of economics. The syllogism "1: RAND work supported the cold war objectives of U.S. policy; 2: Economists worked at RAND; 3: Therefore the work of economists connected to RAND was actually done in service of U.S. cold war objectives" is a mish-mash of political preconceptions and historical confusions. For those seeking to understand the nature of the mathematization of economics, it is vital to recognize that the RAND community had no *disciplinary* commitments. Its role was therefore not one of transforming a discipline but rather diffusing the tools and methods of economic theory into an amalgam of research areas and proto-disciplines that were open to new technical ideas. Mirowski argues otherwise, claiming that "RAND was the primary intellectual influence upon the Cowles Commission in the 1950s, which is tantamount to saying RAND was the inspiration for much of the advanced mathematical formalization of the neoclassical orthodoxy in the immediate postwar period" (2002, 208). At the same time, Mirowski repeatedly and rightly points out that mathematical equilibrium analysis kept economists away from the more eclectic and inclusive approach to social sciences inspired by new computing opportunities—that is, in Mirowski's words, it helped "ward off cyborgs" for decades (see, e.g., Mirowski 2002, 220, 255, 270). If Cowles indeed held off the "cyborg's" debut in economics, it must have been the case that the arcane spirit of mathematical Walrasian economic theory was the villain (or hero). The transformation of economics provided closure to the discipline. It did not embed economics in a transdisciplinary metascience or merge it with management sciences or operations research (as individuals like Anatol Rapoport and Herbert Simon had hoped). Thus RAND was never the dominant force behind the changes in economic theory, as Mirowski and Amadae have claimed. *The inspiration for the integration of economics through the agency of equilibrium analysis came from the mathematics department, not from the military.*

While RAND was not limited by disciplinary concerns, academia was. In the years to come universities, with institutions like departmental instructional units, undergraduate curricula, and Ph.D. program specialization, fostered the differentiation of scientific fields.

The rhetoric of difference between science and engineering, between pure and applied, between literary and scientific disciplines took hold: C. P. Snow's argument in *Two Cultures* against the antidemocratic tendencies of humanities departments emerged in this period. Mathematics departments gained an autonomy they never had before as they became independent from the physical sciences. No longer was mathematics to be taught to undergraduates just so that engineers could build stronger bridges and physicists build bigger bombs. Also at that time the social sciences ruptured with newly differentiated disciplines of business economics (Augier and March 2011), sociology, cultural anthropology, political science, social psychology, and economics (Backhouse and Fontaine 2010) even as they also began to share the same epistemic virtues (Erickson et al. 2013). This is one of the oddities of early cold war science: alongside the differentiation of disciplines, which is still ongoing, there was a shared commitment to certain kinds of techniques among the elites of these disciplines.[8]

Even though many Cowles scholars did work for RAND, there were a number of scholars at Cowles who had no connection with RAND or its activities. Roy Radner, Stanley Reiter, Leo Hurwicz, and Gérard Debreu engaged in mathematical modeling that was more Platonist and less computational. Between Cowles and RAND there was Bourbakian aloofness in the person of the unknowing Debreu. As he asserted later, somewhat ambiguously: "Some of the mathematical economists I knew spent a significant part of the summer at RAND. I did not do that and that may be due to some extent, but not entirely, because I was not a U.S. citizen, and RAND was doing a number of things for the army. . . . I do not know who from

8 The so-called Macy Conferences exemplify this oddity (see Pias 2003/2004). Between 1946 and 1953 these meetings typified the scientific optimism in the postwar U.S. government, and university scientists from different disciplines—psychiatrists, mathematicians, physicists, information scientists, biologists, anthropologists, sociologists, neuroscientists, and economists—worked together to design a society that would resist totalitarianism and prosper in freedom. Leading scientists across all the involved fields discussed topics such as artificial intelligence, neural networks, automatons, and self-regulation, among others. The list of participants is impressive: the psychiatrist William Ross Ashby, the anthropologist Gregory Bateson, the computer engineer Julian Bigelow, the physicist Heinz von Förster, the sociologist Paul Lazarsfeld, the anthropologist Margaret Mead, the social psychologist Kurt Lewin, and the statistician Leonard J. Savage among many others. All of them were surely advocates for redirecting their disciplines toward both scientific (mathematical) rigor and social engineering.

Cowles went to RAND in the summer" (Debreu, in Weintraub 2002, 143–45). Some went, Debreu knew. But he preferred not to ask who went: Who knows what they do there? It's better not to ask. RAND and Cowles, apparently, represented two separate sets of scientific virtues.

## The Cowles Commission

The Cowles Commission for Research in Economics was founded in 1932, two years after the foundation of the Econometric Society. It was named after its creator, Alfred Cowles, a wealthy investment advisor who hoped for better predictions of stock market behavior by using mathematical and statistical tools. For this purpose he hired well-known economists such as Irving Fisher, Harold Hotelling, and Ragnar Frisch to work part time, or as consultants, to the commission. Even though the research produced did not help his business affairs, he continued as the patron of many of the members of the Econometric Society. In 1939, in order to avoid state taxes in the commission's home base in Colorado, Cowles searched for a new home and found it at the University of Chicago.[9]

The move was fortuitous. Chicago had been important in statistical economics, and Henry Schultz, one of the founding members of the Econometric Society, had attempted to unite theory and statistics in his work on estimating demand curves. Chicago's willingness to host Cowles was in part related to the fact that Schultz had recently died in an automobile accident. From the department of economics, only Oskar Lange was part of the Cowles group. Theodore Yntema, a former student of Schultz's, became research director in 1939. As he became more involved in war-related work over the next few years, some of his duties seem to have been handled by Lange: for example, he was the person who wrote to Samuelson and others in 1940–41 asking about possible recruits for Cowles. Instead of seeking local faculty to hire, the Cowles Commission recruited from the pre–World War II European émigré community. As Roy Epstein wrote in his engaging history of econometrics: "It is also appropriate to record

9 The Cowles website (http://cowles.econ.yale.edu) has links to a number of useful histories of the commission.

Cowles's sponsorship of refugees from Nazism, in particular Abraham Wald and Horst Mendershausen. Perhaps owing to the liberal and internationalist outlook of the Cowles family, the Commission soon became a notable stopover point for many foreign economists visiting the United States" (Epstein 1987, 60–61).[10] Jacob Marschak had grown up in the Ukraine, was educated in Germany, had headed the Oxford Statistics Institute, and emigrated to the United States via the New School for Social Research. From Marschak's arrival at Cowles as research director (replacing Yntema) in 1943 the commission became increasingly attractive to technically trained European scholars, most of them well-known to Marschak.[11] Herbert Scarf recalled:

> Marschak was a scholar of great intellectual force, curiosity, and initiative. As director he continued the program of summer conferences, but now there was a dramatic increase in the number of visitors and the size of the resident staff. . . . Leonid Hurwicz had been recruited by Yntema, and in the next several years Trygve Haavelmo, Koopmans, Herman Rubin, Lawrence Klein, Theodore Anderson, Kenneth J. Arrow, Herman Chernoff, Herbert Simon, and other distinguished statisticians and economists were to be associated with the Commission in one way or another. The early research agenda, set by Marschak, was primarily concerned with the particular statistical problems arising in the estimation of parameters in the set of simultaneous equations. . . . Koopmans became the acknowledged leader of that school of econometrics, focusing on the problem

10 It cannot be emphasized strongly enough that the 1930s were a time of both overt and covert anti-Semitism in American higher education. The attempts by some academics, and foundations like Rockefeller, to sponsor and place European mostly Jewish refugee scholars is a well-told story. See, e.g., Feuer 1982; Hollinger 1996; Lipset and Ladd 1971; Lyman 1994; Scherer 2000. For a specific study of how this "worked" in economics, see Weintraub 2014.

11 Cowles's turn toward theory and its concomitant collectivist culture was certainly prefigured under Marschak, specifically in the form of a top-down approach to econometrics. It was also Marschak who, as a real novelty in the organization of research groups in economics, launched biweekly research seminars running concurrently with the summer conferences at Cowles. One might argue that the Activity Analysis Conference reinforced what began with Marschak by adding a noneconometric Walrasian dimension in a context of programming (see Epstein 1987).

of simultaneity and insisting on a complete probabilistic model of the data to be analyzed. (1995, 277)

The empirical focus of the commission was not to last for very long. Lawrence Klein had been recruited by Marschak to join the Cowles staff in order to build a forecasting model of the U.S. economy.[12] This work, which began for him the activity for which he was later to win a Nobel Prize, was very data intensive and required a lot of computation assistance. But because Klein's appointment was not permanent—he was there on a three-year contract—Tjalling Koopmans, Marschak's successor, needed to replace him. After asking Samuelson for some possible candidates (Koopmans to Samuelson, August 16, 1948), Koopmans received Samuelson's reply of August 27 saying that "I have racked my brain but there is no one I can think of to suggest for the opening described in your letter. Good people remain hard to find" (PASP, 45, Koopmans). Consequently more and more of Koopmans's attention was drawn away from empirical modeling to a variety of theoretical problems central to economic policy.

More scientific workers became available after demobilization. The wartime years had seen changes in how economists could join forces with statisticians, mathematicians, and managers in what was coming to be called operations research. In 1948, with RAND focused on national security issues and game theoretic techniques, and with MIT just beginning to follow Samuelson's lead in constructing a research faculty, Cowles was the most important site for reconstructing economic theory as a mathematical science. As a few years earlier Cowles had successfully reconstructed econometric analysis with Haavelmo's work, the adaptation of new mathematical tools was de-

---

12 Klein's thesis advisor, Paul Samuelson (Klein was Samuelson's first Ph.D. student), did not want Klein to take that Cowles appointment, wishing instead to have him remain at MIT to continue his work in macroeconomic theory. Klein, however, was also stubborn. His very left-wing views at the time, not so uncommon among American scholars, found a sympathetic audience among the European social democrats at Cowles. Indeed, Marschak himself, in his Menshevik days, had been the minister of labor in the Soviet Republic of Terek in 1918. Klein's interest in central planning was the reason he jumped at the opportunity to produce a Tinbergen type multiple equation forecasting model of the U.S. economy. In one provocative letter to Samuelson at that time Klein said that he was finally reading all of Marx's *Das Capital* (*sic*) (PASP, 45, Klein).

cisive for economic theory, too. It prefigured the contributions of Arrow, Debreu, and McKenzie. Understanding the context of their work requires understanding the context of this shift to theory at Cowles.

This new postwar era at Cowles became apparent following the withdrawal of one of its major funding sources, the SSRC, as well as, indirectly, the National Bureau for Economic Research (NBER) and the Rockefeller Foundation. The NBER was known for its pioneering use of national accounting and for its cadre of American institutionalists like Wesley Claire Mitchell and Arthur Burns. The more the Cowles program moved away from Schultz's program of empirically deriving demand curves, the greater was the distance between the two research centers. The NBER was skeptical about the theoretical approach to statistical research represented by Klein. That skepticism found voice in a "public" debate occasioned by Koopmans's book review of Burns and Mitchell's 1946 NBER volume, *Measuring Business Cycles*. Koopmans's essay (1947) was titled "Measurement without Theory." This was a rare public encounter between the Cowles research program and other programmatic answers to the question "How should we do economics?"[13] And it was characteristic of Koopmans to take the lead in defending the Cowles program in methodological terms—for it would be Koopmans who acted as the spokesman of the emerging Cowles program.

A second problem Koopmans had to deal with was the growing distance between Cowles and the economics department at Chicago.[14] One of Schultz's and later Hotelling's Ph.D. students, Milton Friedman, was outspoken in his distaste for the theoretical approach to econometrics—a distaste that he expressed clearly in his methodological essay of 1953. This conflict might have appeared to be technical in nature, but it was nourished by a climate of political suspicion. Chicago had grown into a "school" known, among American

---

13 Another confrontation that left its mark on the profession at large was internal to the Econometric Society, as we will see in chapter 7. It remains an open question how this debate echoed the debates about national science policy in general in which Mitchell was involved (Solovey 2013).

14 A recent study by Ross Emmett on the Chicago economics department's workshops provides a comprehensive and fascinating exploration of the relation between Cowles and the department in the 1940s (2011, 97ff).

economists, as staunchly opposed to the Roosevelt administration and New Deal policies (see Horn et al., eds. 2011). Paul Douglas was a liberal outsider among individuals like Henry Simons, Jacob Viner, and Frank Knight, and their students like Gregg Lewis and George Stigler.[15] The Cowles people in contrast were a collection of European socialists and social democrats, and homegrown left-liberals: "[W]e members of the Cowles Commission were seeking an objective that would permit state intervention and guidance for economic policy, and this approach was eschewed by both the National Bureau and the Chicago School" (Klein 1991, 112).

The fact that economists interested in advancing technical tools in economic theory, specifically Europeans, were rather "left" should not have surprised anyone at Chicago. Having fled Nazi and fascist Europe, many Cowlesmen (there were indeed no women) had been active in what was known as the "socialist calculation debate." Oskar Lange, who would resign his professorship at Chicago in 1945 in order to help plan and build the communist postwar economy in Poland, had himself brought forward the Walrasian model as a planning device in his 1936 volume *On the Economic Theory of Socialism*.[16] Thus underneath the surface interest in economic theory was an older set of arguments among economists and social philosophers about the possibility that a centralized planning system, or some kind of socialist planning model, could produce the same efficient outcomes that a competitive market economy might produce. As the interwar versions of this "socialist calculation debate" took shape between market socialists like Lange and Austrian liberals like Hayek, they reflected less on the Russian Revolution that had made the controversy a real

---

15 The conflict had also roots in the United States. Early on in the Cowles time at Chicago the commission took on a project for the wartime Roosevelt administration of examining how rationing could be implemented. By the war's end Milton Friedman and George Stigler wrote what could be thought of as a reply to such thinking called "Roofs or Ceilings?" Published by the right-wing Foundation for Economic Education, this polemic essay against rent controls was paid for and distributed by a national organization of home builders. It was, too, a partial response to the contract work Cowles had taken on to study price controls.

16 Lange was very involved with the wartime Polish government in exile in London and New York. He was admired by Stalin, who persuaded Roosevelt to allow Lange to visit him in the Soviet Union during the war. Thereafter Lange broke with the government in exile, affiliating instead with the communist Lublin Committee. After the war, renouncing his U.S. citizenship, he became Poland's ambassador to the United States.

one and more on the large number of interrelated political positions that were agitating interwar Vienna.

The comprehensive history of these matters in Robert Leonard's history of interwar Vienna (2010) permits our own treatment to be brief. The views of socialists like Otto Neurath and anti-Socialists like Friedrich Hayek were only two of the many facets of the argument. The original statement of the central problem went back at least to Enrico Barone (1908), an Italian economist associated with Vilfredo Pareto, who pointed out that the equilibrium prices were "solved" by the market supply and demand equations: if there were as many equations as unknowns, a solution was assured, and that equilibrium solution was descriptive for any economy, market driven or socialist. Thus the market process could in principle be "found" either through market activity or by the calculation of a planner who had access to the supply and demand relationships. In principle, a centrally planned economy could replicate the allocative efficiency of a market economy. Issues of social welfare took on new meanings in this debate.

These controversies had a real bite to them in the interwar period, as the early planning experiments in Eastern Europe (e.g., Béla Kun in Hungary) suggested that the philosophical problems were not unrelated to the practical problems facing the planner: issues of prices as information "carriers" were newly important, and the problem of planning's practical side appeared to require analytic methods that had not previously required solutions. Each of our protagonists grew up with these discussions. Arrow recalled:

> On returning from military service, I planned to write a dissertation which would redo *Value and Capital* properly, a very foolish idea. I had two motivations. One was to supply a theoretical model as a basis for econometric estimation. The other was a strong interest in planning. I would have described myself as a socialist, although one that had a strong belief in the usefulness of markets. Market socialism was a widespread view. Hotelling held it. It had been popularized especially by the works of O. Lange (reprinted in Lipincott 1938) and A. P. Lerner (1946). In the immediate postwar period, the idea of national planning

> to supplement markets was common in Western Europe, and allocation in effect was treated, in principle, as the solution of a general equilibrium system (although with many simplifications). It appears in retrospect that the planning had little effect (good or bad) on the development of the European economies, but a great deal of intellectual energy was expended. (2009, 7)[17]

Most economists understood that a socialist model provided the clearest vision of the role of markets in allocating resources. Since the allocation problem was revealed cleanly in such a system, discussions of market socialism were common in the 1940s. Nearly everyone discussing a theory of production in the second half of the 1940s in Chicago invoked Oskar Lange's theory of planning.

In addition to the theoretical problems caused by socialist institutions, we noted earlier that during the war the U.S. economy had necessarily become a centrally planned economy. It was not the case that the U.S. economy stood at one end of a continuum with respect to planning with the USSR at the other end. The United States did not achieve victory in the war by letting free or competitive markets decide what armaments to produce and how scientists should be allocated to particular military tasks. In fact, during the war new kinds of techniques for allocating resources had come into use. It was out of such work that Cowles would make its greatest theoretical contributions, and Koopmans's activity analysis, as we will see in the next chapter, was one of them.

The spirit of Cowles, to anticipate that discussion, was not political in nature. The tacit agreement among members of the staff was rather that the significance of technical tools derived in some way or other from the possibility of social engineering—*whatever* might be the political or moral justification for doing so. That is, applying new tools to economic theory, Koopmans's research program represented

---

17 In her extensive exploration of the intercalated emergence of neoclassical economics and market socialism, Johanna Bockman noted that "To Oskar Lange [writing in the 1930s], only a socialist system could attain free competition as described by economists and could maximize social welfare because it did away with private property in those areas that lack competition and thus removed the obstacles to free competition. Lange soon afterward moved to the United States and began to teach neoclassical economics at the University of Chicago" (Bockman 2011, 19).

a *depolitization* of economic theory. Otherwise, Koopmans would have fallen under the suspicion not only of the economics department but of the university's administrators and government officialdom at large. In that period in which interest in socialism was tantamount to sedition, past Communist Party membership was grounds for employment termination, and interests in economic planning were best left unremarked, the need for depolitization was evident to Koopmans. As Herbert Simon commented on these years: "By 1948, Communists and supposed Communists were being discovered under every rug. . . . Any graduate of the University of Chicago, with its reputation for tolerance for campus radicals, was guaranteed a full field investigation before he could obtain a security clearance" (Simon, in Mirowski 2002, 246).

Koopmans's shift to greater abstraction in economic theory was crucial for utilizing the planning tools developed during wartime research and reshaping their previous meanings to the new environment. Rather inconspicuously wartime social engineering shed any association with "socialism" and instead emerged as a project that secured Western liberty. During the politically tense years after World War II, there were few economists with a background in the economic theory of socialism who would not have wanted to embrace Koopmans's policy since economic theorems were unlikely to be interrogated by the House Un-American Activities Committee.

In this transition period, the Cowles Commission behaved more like postwar mathematics departments than economics departments. The self-protective depolitization of mathematicians who had been active in wartime military projects appealed to the economists at the Cowles Commission. Mathematical purity was especially attractive to the mathematics departments at Princeton and Chicago. Both had been involved in the recruitment of mathematicians during World War II, and both would become central to the postwar institutions of science. We contend that the Cowles Commission and this mathematics community mutually stabilized one another as they justified the relevance and authority of each other. Thus we will need to understand the important role of Chicago and Princeton mathematicians for most Cowlesmen in order to understand the transformation of economic theory that was birthed there.

## Mathematics at Chicago

It is a commonplace to say that the success of the natural sciences through the end of the nineteenth century depended on advances in mathematics (Warwick 2003; Volterra 1906). However, mathematics could not rise to *primus inter pares* among the sciences simply by providing tools for the special sciences. It also had to strengthen its *disciplinary autarky*. Mathematics could only flourish in the postwar years if it stood apart from the applied disciplines whose political commitments produced continual scrutiny in that McCarthy period. If all mathematics were applied mathematics, and applications bore political weight in the cold war era, mathematicians would live under the same security regime as did nuclear physicists. This is the background of the emergent and valorized distinction between "pure" and "applied" mathematics: the latter was potentially contaminated by the applications that existed outside mathematics, while the former entailed no commitment to any ideas that were not mathematical.[18] This relief from responsibility associated with mathematical purism was openly assumed by Jean Dieudonné (as Bourbaki), whose mathematics, as we have seen, provided shelter also to students at École Normale living in an occupied Paris.

> Why do applications [of mathematics] ever succeed? Why is a certain amount of logical reasoning occasionally helpful in practical life? Why have some of the most intricate theories in mathematics become an indispensable tool to the modern physicist, to the engineer, and to the manufacturer of atom-bombs? Fortunately for us, the mathematician does not feel called upon to answer such questions. (Bourbaki 1949, 2)

This Bourbakian aloofness contributed to its success in the United States. As German mathematics had self-destructed in the 1930s with

18 The therapeutic efficacy of mathematics was of course no novelty. The Cambridge mathematician G. H. Hardy, whose book *Pure Mathematics* was a major success after its initial publication in 1908, wrote about these matters in a poignant essay *A Mathematician's Apology*. Published in 1940 with the encouragement of C. P. Snow to help Hardy overcome his profound depression over the start of World War II, the essay allowed Hardy to say that, as a serious mathematician, "I have never done anything 'useful'. No discovery of mine has made, or is likely to make, directly or indirectly, for good or ill, the least difference to the amenity of the world" (Hardy [1940] 1969, 150).

the purges of Jewish scholars, and with American mathematics still backward compared to European scholarship, Bourbaki represented a new and exciting integrative vision of the discipline of mathematics, one that took hold internationally in the period from the end of the 1940s through the late 1960s. The *Theory of Sets* volume had appeared in 1939, but the war's dislocations meant that subsequent volumes only began appearing in 1947. Those new volumes took up topology, algebra, functions of one real variable, and integration. Bourbaki mathematics became metonymous for pure mathematics, particularly at Chicago.

The story of the reconstruction of the Chicago mathematics department begins with Marshall Stone. Having done classified research for the U.S. Navy and the Department of War until 1945, he was asked to rebuild Chicago's mathematics department in 1946. In a reminiscence of that period, he wrote, "In 1946 I moved to the University of Chicago. An important reason for this move was the opportunity to participate in the rehabilitation of a mathematics department that had once had a brilliant role in American mathematics but had suffered a decline, accelerated by World War II" (1989, 183). Saunders Mac Lane, who would become a major figure in U.S. mathematics in the following decades, was hired by Stone in 1947 and recalled that period:

> Robert Maynard Hutchins, president of the University of Chicago (1929–1951), had brought the Manhattan project to the university during WWII and with it many notable scientists including Enrico Fermi, James Franck, and Harold Urey. As the war drew to a close, he and his advisors decided to try and hold these men and their associates at Chicago. . . . [They] realized that this should be [the occasion for] a much-needed strengthening of the department of mathematics. With the advice of John von Neumann (who had been associated with the Manhattan project), they approached Marshall H. Stone, then a professor at Harvard, suggesting (after some talk of a deanship) that he come to Chicago as chairman of mathematics. . . . After receiving suitable assurances from President Hutchins, Stone

> came to Chicago in 1946. He thereupon brought together what was in effect a whole new department. (Mac Lane 1989, 146)

As the Chicago mathematics department's own online history notes, "There were at the time five vacant senior positions which had accumulated during the war, which meant that the department had to be rebuilt almost completely, and there was a wish to match the level of appointments in the physical sciences which the university had been able to make through its involvement in the Manhattan Project" (ibid.).[19] Stone was a magnificent administrator who knew that to pilot mathematics into new postwar waters he should take advantage of the opportunity to recruit first-rate émigré mathematicians. As full professor he hired Antoni Zygmund (originally from Poland), Shiing-Shen Chern (originally from China), as well as the American Saunders Mac Lane from Harvard, all of whom arrived in 1947. As assistant professors he hired Irving Siegel, Edwin Spanier, and Paul Halmos (originally from Hungary), who was the former assistant of von Neumann at the Institute for Advanced Study. Irving Kaplansky and Abraham A. Albert remained of the old faculty and were immediately drawn into the new spirit that was set by the single most important hire in the department, André Weil. We noted earlier Debreu's and McKenzie's excitement about this new department; McKenzie took classes from Kaplansky, Halmos, and Mac Lane. Mac Lane himself recalled: "In this period at Chicago, there was a ferment of ideas, stimulated by the newly assembled faculty and reflected in the development of the remarkable group of students who came to Chicago to study. Reports of this excitement came to other universities; often students came after hearing such reports" (ibid., 148).

André Weil's autobiography (1991) focuses on the period up until 1947 when he arrived at Chicago. An important figure in Bourbaki, Weil as a Jew had to leave France after the German invasion, ending up in the United States. After securing a Rockefeller-funded position first at the New School then at Haverford College, Weil spent an intellectually unrewarding period at Lehigh University in Bethlehem,

19 http://math.uchicago.edu/about/history.shtml.

Pennsylvania.[20] Facing a variety of complications due to his refugee status, he was fortunate to get a visa to leave the United States to take up a position at the University of Saõ Paulo in Brazil. As he recalled:

> As soon as Paris was liberated in 1944, mail service started up again . . . and I resumed my correspondence with Bourbaki. Before the liberation, I had received a report on only one congress: this was the miniconference held in September, 1943 . . . attended by Cartan, Delsarte, and Dieudonné. . . . The problem arose of sending the conclusions to Chevalley and me. . . . [A copy] was sent to London by clandestine Gaullist mail pouch, [and] somehow reached New York, by which time it had lost its address. By mere chance it was seen by the physicist Francis Perrin, who thought he recognized Bourbaki's style and sent it straight away to me. Once Paris was liberated and I was in Brazil, such detours became unnecessary, and our discussions resumed as usual. (Weil 1991, 188)

Nonetheless, it is clear that from the time he arrived in the United States in March 1941 Weil was concerned to maintain his own mathematical relevance. He was indefatigable in reestablishing connections with other French mathematicians who had also managed to escape Europe and arrive in the United States and who shared his Bourbakian vision. Connections with his friends Irving Siegel and Claude Chevalley were reestablished as Chevalley had duly reported to the French Consulate in New York and was told to stay on in Princeton where Lefschetz had secured him an appointment as assistant professor (Weil 1991, 178).

---

20 He described this period in terms quite revealing of the Bourbakist aloofness that he imported to the United States: "The institution to which I belonged (a word that all too accurately describes my relationship with my employer) was graced with the noble title of 'university'; but in fact, it was only a second-rate engineering school attached to Bethlehem Steel. The only thing expected of me and my colleagues—who were totally ignorant as far as mathematics went—was to serve up predigested formulae from stupid textbooks and to keep the cogs of this diploma factory turning smoothly. Sometimes, forgetting where I was, I would get carried away and launch into a proof. Afterwards, according to the well-established ritual, I would always ask, 'Are there any questions?' Just as predictably would come the question: 'Is that going to be on the exam?' My answer was ready: 'You should know it, but it's not very important.' Everybody was happy. . . . At least I was able to continue writing, though at a somewhat slower pace, my book on the foundations of algebraic geometry, the indispensible key to my later work" (180–81).

Having hired Weil, Marshall Stone had surely strengthened pure mathematics. But he also had to place applied mathematics somewhere. In this respect the presence of the Cowles Commission had an impact on the role of applied mathematics at Chicago:

> During my correspondence of [19]45–46 with the Chicago administration I had insisted that applied mathematics should be a concern of the department, and I had outlined plans for expanding the department by adding four positions of professors of applied subjects. . . . Circumstances were unfavorable. . . . On the other hand, there was pressure for the creation of the Department of Statistics, exerted particularly by the economists of the Cowles Foundation. A committee was appointed to make recommendations to the administration for the future of statistics with Professor Allen Wallis, Professor Tjalling Koopmans, and myself as members. Its report [urged] the creation of a Committee on Statistics, Mr. Hutchins being firmly opposed to the proliferation of departments. The committee enjoyed powers of appointment and eventually of recommendation for higher degrees. . . . and developed informal ties with the Department of Mathematics. (Stone 1989, 188–89)[21]

Thus the presence of the Cowles Commission diminished pressure on the mathematics department to integrate applied mathematics into its curriculum. This is an additional reason why the ties of the Cowles Commission to the mathematics department were tighter than to the economics department and helps explain why few at Cowles were deeply involved in the controversies energizing the larger economics profession. Cowles and the mathematics department were in a mutually stabilizing relationship: Cowles allowed the mathematics department to avoid issues of irrelevance while connection to the mathematics department rather than the economics department allowed Cowles to avoid problems of political advocacy.

---

21 Leonard Savage also played a role in the foundation of the statistics department in 1949. A student of von Neumann and Marston Morse at Princeton, he was hired as a research associate in 1947. He was the only one of the "high-tech" scholars at Chicago working jointly with the economics department, specifically with Milton Friedman (Friedman and Savage 1948).

The connections between the Chicago mathematics department and the Cowles group were also manifest in the work of several mathematicians. One of them was Murray Gerstenhaber. A Ph.D. student of Abraham Albert, he would contribute to the conference on activity analysis that we describe in detail in the following chapter. Another was Israel N. Herstein, who would write the important textbook *Topics in Algebra*. Herstein was connected to the Cowles group after his 1952 appointment as assistant professor at Chicago in mathematics and economics. In mathematics he worked closely with Albert, and he also wrote several Cowles Commission Discussion Papers on optimization concepts and efficiency. Herstein was, next to André Weil, one of the early mathematical confidants of Debreu. He also became Debreu's first coauthor, producing "Nonnegative Square Matrices" (Debreu and Herstein 1953). That same year Herstein wrote a paper jointly with the Princeton and RAND mathematician (and future Fields Medalist) John Milnor, which must have been a blueprint for the style Debreu wished to develop (Herstein and Milnor 1953). The opening footnote thanked just one person, Debreu, for his comments.

Morton L. Slater was another mathematician affiliated with Cowles as a "research consultant" who became a key figure for both Debreu and McKenzie. Slater was a University of Wisconsin graduate who did his graduate work in mathematics at Harvard. He served as the referee of most of the papers at the 1949 conference that we discuss in the next chapter. In that period he also had the title of Senior Mathematician, (Navy) Ordinance Research, Chicago, Illinois.[22] In Cowles's 19th Annual Report for July 1, 1950–June 30, 1951, titled "Rational Decision-Making and Economic Behavior," Slater is described as a research associate who

> studied the mathematical content of the problem of economic choice, that is, the problem of maximization under constraint. He provided mathematical advice and criticism with regard

22 As mentioned in the "Annals of the Computation Laboratory of Harvard University Volume XXVI," which reported on the Proceedings of a Second Symposium on Large-Scale Digital Calculating Machinery, September 13–16, 1949.

> to the work of many staff members and provided expository presentations of mathematical results to economists. Visiting mathematicians addressed the staff on problems of maximization under linear inequalities. John Chipman, Debreu, and Slater explored mathematical theorems providing criteria of stability in models of international, interregional, or interindustrial economics. (Cowles Report 1951)

There was one decisive working paper by Slater in 1950—"Lagrange Multipliers Revisited"—that made both Debreu and McKenzie aware of the use of Shizuo Kakutani's version of the fixed-point theorem that would be central for their simultaneous existence proofs. Each separately recalled that they learned of Kakutani, and von Neumann's first use of the same theorem, by reading Slater. Slater, two years younger than Debreu, seemed headed for a very similar academic career as Debreu, so it was appropriate for Koopmans to let them share the same office.

There is a final shared mathematical influence on Debreu's and McKenzie's existence theorem work: they each recalled that they saw a notice on the mathematics department's bulletin board advertising the availability of Marston Morse's lecture notes on convex sets. Morse, a mathematician at the Institute for Advanced Study in Princeton, was known for his work on differential topology. Both Debreu and McKenzie got acquainted with convexity analysis from those mimeographed notes.

### John von Neumann

With the emigration of Albert Einstein, Kurt Gödel, Hermann Weyl, and John von Neumann in the early 1930s, Göttingen and Vienna were brought to the Institute for Advanced Study in Princeton. When Oswald Veblen and James Alexander moved to the institute from the university, the institute became the center of the mathematical universe. Among all the institute and university mathematicians, *primus inter pares* of wartime engineering, operations research, and game theory was John von Neumann. While André Weil, the Chicago representative of Bourbaki, was to personify the new intellectual purity imported into economic theory via Gérard Debreu and others, at Princeton John von Neumann's authority fused pure mathematics with the eclectic spirit of applied research. The work of McKenzie, Arrow, and Debreu would differently make manifest this fusion.

Recent historical work has placed von Neumann in the intellectual and government communities of the early cold war era. In an unusual historiographic confluence, the long period in which von Neumann was either ignored, castigated as a war-monger like Kubrick's Dr. Strangelove (Heims 1980), or treated as a genius to be worshiped (Macrae 1992) has been followed by serious historical studies of von Neumann's complex mathematical and personal history and his profound importance not only to U.S. military efforts during World War II but to atomic energy, atomic weaponry, and computers in the postwar period (Asprey 1990; Mirowski 2002; Israel and Gasca 2009; Leonard 2010). "Our" von Neumann here is the von Neumann whose work grounded, unwittingly perhaps, the simultaneous contributions of Arrow, Debreu, and McKenzie.

While economists often claim von Neumann as one of their own, he was always a mathematician: in his six-volume *Collected Works*, the first hundred pages of the five-hundred-page volume 6 contain all his papers on economic equilibria and the theory of games. Both as a mathematician and wartime consultant on numerous secret projects, von Neumann was akin to the old-fashioned long-distance telephone operator of the 1930s: he coordinated the many conversations that intersected his broad research in mathematics and its "applications." During the war, when rail not air travel was the norm, he was frequently in Chicago where he had to change trains from Princeton (via Philadelphia) to Albuquerque, New Mexico (and Los Alamos). Breaking his trip in that city, he often met with the physicists and mathematicians at the University of Chicago and relayed to them developments in several different fields of mathematics and nuclear science (see Mirowski 2002). His connection with people at Cowles, mathematicians and statisticians, was thus indirect although his presence was noted.

As we will shortly see when we examine the existence proofs of the early 1950s, some of von Neumann's own earlier work was foundational for economic analysis. His article "Über ein ökonomisches Gleichungssystem und eine Verallgemeinerung des Brouwerschen

Fixpunktsatzes" (On an economic system of equations and a generalization of the Brouwer fixed-point theorem) had been published in 1937. He had presented it in 1934 in Karl Menger's mathematical seminar in Vienna and again in the academic year 1939–40 in Princeton in English with the title "A Model of General Economic Equilibrium." In the next chapter we will see how the activity analysis conference "resurrected" von Neumann's paper and linked it to programming and game theory. That paper examined the possibility of the existence of equilibrium of a model economy whose various interconnected parts exhibited uniform rates of growth. It was the first paper related to economics that used a fixed-point argument to achieve an equilibrium solution. Arrow's, Debreu's, and McKenzie's simultaneous proofs of equilibrium in a competitive model economy were a direct result of the rediscovery of the 1937 von Neumann paper in the late 1940s at Koopmans's Cowles.

McKenzie, from his short time as graduate student in Princeton between 1939 and 1941, had heard von Neumann present his paper to the Princeton economics department. Years later, on November 15, 1976, he wrote to Oskar Morgenstern saying: "You may not recall that I was present when von Neumann spoke to the economics seminar on his growth theory. I was among those whose incomprehension distressed you. I recall questioning the use of mathematics in economics and drawing the response from von Neumann that this was just the way physics was viewed before Newton, and that a new type of mathematics would be needed for economics" (OMP, 52, Neumann).[23]

In his 1944 game theory book with Morgenstern, von Neumann was clear that the "calculus" was inappropriate for economic analysis. His growth model in fact employed a topological method, namely the Brouwer fixed-point theorem. "The connection with topology," von Neumann wrote, "may be surprising at first, but the author thinks that it is natural in problems of this kind" (1945 [1937], 1). Von Neumann himself did not inquire into this "natural connection." By the late 1940s he began pursuing algorithmics and the computational

23 The authors are grateful to Teresa Tomas Rangil for locating this letter in the Morgenstern papers at Duke University.

use of mathematics, eschewing the Bourbakist "mother-structure" of mathematics, topology. His 1947 paper "The Mathematician" stated clearly his view that difficult problems in science generated the most important mathematical discoveries and that a mathematics dependent only upon itself would soon become uninteresting, even to mathematicians. This view was entirely anti-Bourbaki.

The resurrection of von Neumann's growth model poses a second irony with respect to his aversion to competitive economic analysis. The central theorem of that paper is related to his 1928 proof of the minimax theorem for two-person zero-sum games. While in that paper (von Neumann 1928) he gave no reference to any economic tradition, the later growth model referred to a "typical economic equation system" ([1937] 1945, 1). While the former modeled strategic behavior, the latter modeled competition. In the 1944 *Theory of Games and Economic Behavior* strategic behavior and competition were expressly opposed to one other (1953 [1944], 15). And yet the two papers' central theorems were "oddly connected," von Neumann wrote, for being reducible one to one another via a saddle point (1945 [1937], 1), a point that was simultaneously a maximum and a minimum. Von Neumann and Morgenstern wondered whether "there may be some deeper formal connections here. . . . The subject should be clarified further" (1953 [1944], 154). Without anyone clarifying the subject further, the equivalence eliminated the possibility that game theory could develop as an alternative "paradigm" to competition. The different economic intuitions that had separately evolved in game theory and in perfect competition analysis merged since they were both amenable to the same theorems.

In a footnote to that discussion on page 154 of the *Theory of Games and Economic Behavior* (1944), von Neumann (and it was certainly von Neumann) pointed out that the existence theorem for two-person noncooperative games was first proved by him in 1928. He then said that he did the proof in a more general form in his 1937 paper. The footnote ended by pointing out that in 1941 Kakutani further extended that fixed-point theorem's range. Kakutani (1941) would function as a catalyst for von Neumann's skepticism about competitive analysis. He generalized von Neumann's fixed-point theorem *without* reference to any economic context and that then would become the standard reference for McKenzie's, Arrow's, and Debreu's proofs. Von Neumann himself did no further work in competitive equilibrium analysis, or any economics, nor did he return to work on topological fixed-point proofs.[24]

These unintended effects of von Neumann's work on Arrow, Debreu, and McKenzie can be better understood when considering von Neumann's direct impact on the work produced at the Princeton mathematics department, for example, John Nash's 1950 paper on equilibrium for n-person games. That paper sparked the idea for proving existence for Arrow, who, in contrast to Debreu, would remain true to its origins in modeling strategic behavior. But the main influence of von Neumann's work on game theory at Princeton was in the use of computational methods as represented by his colleague at the mathematics department in Fine Hall, A. W. Tucker.

## Princeton's Fine Hall

Unlike Chicago, Princeton's mathematics department hardly needed strengthening. Veblen, who had been Princeton's department chair, helped leadership pass to Solomon Lefschetz,[25] who was soon to recruit a remarkable faculty by replacing senior retirees with young stars like Alonzo Church, William Feller, A. W. Tucker, S. S. Wilks, and Emil Artin. In contrast to the unified mathematical spirit developed by Stone and Weil at Chicago, the mathematical hothouse of Fine Hall, home to the Princeton University mathematics department, hosted an assemblage of groups from which a hierarchy emerged: at the top was Lefschetz, who jointly with Ralph R. Fox and Norman Steenrod represented topology (just as at Chicago). Steenrod and the Bourbakian Samuel Eilenberg worked on homology theory. Fox's most important student was John Milnor, who, even before concluding his Ph.D. on link groups, had collaborated with Israel Herstein on an axiomatic approach to utility theory (1953). Separate from the Lefschetz students, there was Salomon Bochner

---

24 Morgenstern (1941) contributed his own views when he discussed von Neumann's article together with Wald's when reviewing Hicks's *Value and Capital* (1939).

25 Mac Lane (1989b, 220) recalled the Princeton ditty about Lefschetz: "Here's to Lefschetz, Solomon L. / Irrepressible as hell / When he's at last beneath the sod / He'll then begin to heckle God."

working in analysis. Herbert Scarf, who would come to be known for the computational use of the fixed-point theorem in general equilibrium analysis, was supervised by Bochner. Then there was a group around Emil Artin in algebra with his students John Tate and Serge Lang, who would both become leading Bourbakists. In this spirited surrounding Albert William Tucker advanced von Neumann's game theory. Tucker, a former student of Lefschetz's, supervised virtually all the future top game theory students: David Gale (Ph.D. 1949), John Nash (Ph.D. 1950), Lloyd Shapley (Ph.D. 1954), and Marvin Minsky (Ph.D. 1954).[26] Though supervised by Fox, Harold Kuhn belonged to this group, too, as well as Leon Henkin. Gale, Kuhn, and Tucker ran a weekly seminar on game theory and its computational uses.

As the Cowles Commission had little contact with the Chicago economics department, so too was Tucker's group uninvolved with the Princeton economics department. Oskar Morgenstern was the lone representative of recent developments in economic theory though his students rather shunned economic theory. As Martin Shubik recalled:

> The graduate students and faculty in the mathematics department interested in game theory were both blissfully unaware of the attitude in the economics department, and even if they had known of it, they would not have cared. . . . The contrast of attitudes between the economics department and the mathematics department was stamped on my mind soon after arriving at Princeton. The former projected an atmosphere of dull business-as-usual conservatism of a middle league conventional Ph.D. factory; there were some stars but no sense of excitement or challenge. The latter was electric with ideas and the sheer joy of the hunt. Psychologically they dwelt on different planets. If a stray ten-year-old with bare feet, no tie, torn blue

26 Scarf also lists others connected to game theory at Princeton: Richard Bellman, Hugh Everett, John Isbell, Samuel Karlin, John Kemeny, John Mayberry, John McCarthy, Harlan Mills, William Mills, Norman Shapiro, Laurie Snell, Gerald Thompson, David Yarmish, Ralph Gomory, and William Lucas (Scarf 1995).

> jeans, and an interesting theorem had walked into Fine Hall at tea time, someone would have listened. When von Neumann gave his seminar on his growth model, with a few exceptions, the serried ranks of Princeton Economics could scarce forbear to yawn. (1992, 152–53)

The group surrounding Tucker was uninterested in relating their research to traditional questions of economic theory. They followed von Neumann's anti-Bourbaki lead in mathematical engineering as they developed game theory as a computational rather than a formal discipline. This perspective allowed the developments in linear programming, game theory, and operations research to merge. The notion of strategic behavior was lost to economics as it moved into the foreground of organizational questions. The *Theory of Games and Economic Behavior* was thus pushed away from the mainstream culture in economics. This was the period in which von Neumann gave a game theory talk to the economics department at MIT and on returning to Princeton told Morgenstern that "Samuelson is no mathematician . . . and even in 30 years he won't absorb game theory" (von Neumann, in Mirowski 2002, 139n41). As Shubik witnessed,

> Unknown to me at the time was the breadth of the activity going on in linear programming. By 1947 von Neumann had conjectured the relationship between the linear programming problem and its dual and the solution of zero-sum two-person games. Gale, Kuhn, and Tucker started to investigate this more formally by 1948 and published their results in 1951. Kuhn and Tucker were also active in the development of nonlinear programming. The seminar at Fine Hall lumped the newly developing mathematics of game theory and programming together. Although there was a beautiful link between the mathematics for the solution of two-person zero-sum games, to a certain extent this link may have hindered rather than helped the spread of game theory understanding as a whole. For many years operations research texts had a perfunctory chapter on game theory observing the link to linear programming and treating linear programming and game theory as though they were one. The

economics texts had nothing or next to nothing on the topic. (1992, 159–60)

While Shubik contrasted von Neumann's game theory with programming, George Dantzig, who discovered linear programming, could equally credit von Neumann as his source.[27] The link is revealed in Robert Dorfman's (1984) particularly lucid account of the discovery of linear programming. Dorfman quoted an unpublished memoir from 1976 by Dantzig as follows:

> On October 1, 1947, I visited von Neumann for the first time at the School [*sic*] for Advanced Study at Princeton. I remember trying to describe to von Neumann the Air Force's problem [of airframe production]. I began with the formulation of the linear programming model in terms of activities and items, etc. Von Neumann did something which (I believe) was not characteristic of him. "Get to the point", he said impatiently. Having a somewhat low kindling-point myself at times, I said to myself, "OK. If he wants a quick version of the problem, then that's what he will get." In under one minute I slapped a geometric and the algebraic version of the problem on the board. Von Neumann stood up and said, "Oh that!" He then proceeded for the next hour and a half to lecture to me on the mathematical theory of linear programs (as it later came to be called). At one point seeing me sitting there with my eyes popping and my mouth open (after all I had searched the literature and found nothing), von Neumann said something like this: "I don't want you to think I am generating this out of my head on the spur of the moment. I have just recently completed a book with Oskar Morgenstern on the theory of games. What I am doing is conjecturing that the two problems are equivalent. The theory that I am outlining to you is really an analog of the one that we have developed for the theory of games." Thus I learned about Farkas's lemma and about duality for the first time. (1984, 291–92)

---

27 See the exceptionally informative interview of Dantzig in Albers et al. 1986.

There followed a series of memoranda back and forth between von Neumann and Dantzig; in one there is a footnote crediting Koopmans for "making suggestions on which the procedure for moving from one simplex to a better one is based" (ibid., 292; see also Backhouse 2012, 6). Dantzig visited Cowles in June 1947, met Koopmans, and discovered the same interests. In June 1948, Dantzig visited Princeton to work with Kuhn and Gale on a solution of the duality problem, from which the Kuhn-Tucker theorem emerged. And so Koopmans, Dantzig, and Kuhn became the central figures in the conference on activity analysis in June 1949, to which we now turn.

# CHAPTER 5
## COMMUNITY

In December 1948, in one of the hotel rooms at the annual meeting of the American Economic Association, several scholars with various backgrounds but similar interests conceived the idea of a conference (Kuhn 2008). Tjalling Koopmans, Harold Kuhn, George Dantzig, Albert Tucker, Oskar Morgenstern, and Wassily Leontief were all involved in projects that had the idea of linear programming in common. Koopmans, who in summer of 1948 had become the research director at the Cowles Commission, would take the major initiative in organizing the conference in the context of a research contract with the RAND Corporation, signed in January 1949, titled "Theory of Resource Allocation." It called for Koopmans to bring together a small group of individuals who had been working on these kinds of projects to a conference titled "Activity Analysis of Production and Allocation," which he scheduled for June 20–24, 1949.

That conference was the "coming-out party" for the community in which Arrow's, Debreu's, and McKenzie's careers would flourish. That conference announced, more than any other single event, the emergence of a new kind of economic theory growing from game theory, operations research, and the related mathematical techniques of convex sets, separating hyperplanes, and fixed-point theory. The central idea of the conference was to develop the theory of linear programming, one of the successful methods of planning that had grown out of wartime research, and extend it to the more general economic theory of (not necessarily wartime) production. What later would be a commonplace for economists was discussed in nuce, in an early state when the boundaries between the theories were not yet drawn, the structures not yet settled. The conference established, as it were, the historical conditions for the possibility that Arrow's, Debreu's, and McKenzie's work would mark a new era in economics.

This new world of theory would not be visible everywhere, even at the important sites of academic economics: Harvard, for example, had made its intentions known when they let Samuelson go to MIT in 1940. Nor would this new work be carried out by the older members of the Econometric Society. Harold Hotelling at Columbia had recently abandoned economics to create the first statistics department in the United States at the University of North Carolina at Chapel Hill, while Griffith Conrad Evans had years earlier turned his attention to building the Berkeley mathematics department. It was left to the Cowles Commission to bring together a new community from the various sites discussed in the previous chapter. Koopmans's introduction to the 1951 proceedings described the eclectic spirit of the conference:

> [In] this conference, scientists classifiable as economists, mathematicians, statisticians, administrators, or combinations thereof, pool their knowledge, experience, and points of view to discuss the theory and practice of efficient utilization of resources. The mathematicians brought new tools of analysis essential to the progress of economics. The administrators introduce an element of closeness to actual operations and decisions not otherwise attainable. Those speaking as statisticians adduce data and discuss their limitation. The economists contributed an awareness of a variety of institutional arrangements that may be utilized to achieve efficient allocation. (1951a, 4)

As important as the conference appears in hindsight, it was not intended or planned as a path-breaking event. Hardly any of the participants could have recognized at the time the effects it would have. Koopmans began planning the event only some months beforehand and had sent out the invitation letters only one month before the conference while circulating drafts of papers just weeks before the meeting. The conference, as we are going to show, was meant to be a gathering of scholars pulling at the same string rather than the rally of a community proclaiming a new future of economic research. The conference would make manifest a new spirit among participants only because they did not quarrel about their different backgrounds, their different intellectual socializations, and the diverse possible consequences of their work. Instead they communally enjoyed their skills while they only implicitly agreed on their meaning.

## Snapshot

In his conference invitation letter to individuals around the United States, Koopmans identified a number of groups representing "economists, mathematicians, statisticians, or combinations thereof": the Department of the Air Force, the Bureau of Labor Statistics, the Council of Economic Advisers, the Cowles Commission, the Harvard Economic Project, the Johns Hopkins Group, the Navy Department, the Princeton Group, the RAND Corporation, and the Stanford Group. The Princeton University group was represented by the mathematicians Albert Tucker, Harold Kuhn, and David Gale (who shortly before had left Princeton's mathematics department for Brown) and the "connected" economists Oskar Morgenstern and Ansley Coale. From Cowles, we find Kenneth Arrow, Murray Gerstenhaber (a mathematics department Ph.D. student of Abraham A. Albert), Clifford Hildreth, Tjalling Koopmans, Stanley Reiter, and Herbert Simon. From RAND or the Department of the Air Force we find Marshall Wood, Charles Hitch, Paul Samuelson (who had been visiting RAND from MIT), Murray Geisler, George Brown, and a key figure of the conference, George Dantzig, who alone gave four papers. Others with different affiliations include Robert Dorfman, Harlan Smith, and Yale Brozen. Nicholas Georgescu-Roegen was present from the Harvard Economic Project. In total, there were thirty-four papers and about fifty participants. Koopmans also invited von Neumann, but he did not come, nor did Leontief, who was ill (see Backhouse 2012).[1]

---

1 Individuals who presented papers that were not incorporated in the volume included Merrill Flood, David Hawkins, Leonid Hurwicz, Abba Lerner, and Marvin Hoffenberg. Among those who attended the conference but did not present papers were Armen Alchian, Evsey Domar, and others from the Cowles Commission like Jean Bronfenbrenner Crocket, George Borts, Carl Klahr, Jacob Marschak, and William Simpson. From Princeton Thomson Whiten and Max Woodbury were in attendance, while individuals like Tibor Scitovsky and Oswald Brownlee were present

What stands out is the interrelatedness of the groups from both government and academic institutions. Marshall Wood, the named representative of the Department of the Air Force, moved back and forth to RAND, which was doing contract work on game theory. Samuelson, the 1947 inaugural Clark Medal winner, was visiting RAND that spring.[2] Data issues confronted the Bureau of Labor Statistics, the Council of Economic Advisers, and Leontief's Harvard Economic Project as well as Cowles, where Lawrence Klein, prior to his leaving for Michigan after a year spent in Oslo with Ragnar Frisch, had been constructing a national econometric forecasting model along lines that Jan Tinbergen had pioneered. Evsey Domar at Johns Hopkins and Tibor Scitovsky at Stanford were studying technological change. The mathematician Mina Rees, representing the navy, had been supporting work in mathematical economics through the Office of Naval Research: Morgenstern at Princeton as well as Arrow at Stanford had been receiving such funds as had von Neumann as a consultant at RAND. It was a very small world.

The theories discussed at the conference likewise stood "between" the concerns of academia and the national laboratories. As Koopmans described the purpose of the meeting to his invitees on May 11, 1949:

> [The conference concerns] a related group of techniques for the analysis and planning of resources allocation, that have become known under the name "linear programming". The problem area involved includes: the inter-industry relations technique developed by W. Leontief, and the related data and

as well. In the large cast of participants we find the names of four future Nobel Prize winners in economics.

2 At RAND, Samuelson was working at Hitch's request on a monograph that would explain how linear programming might be improved by economic theory, which years later would evolve into Dorfman, Samuelson, and Solow 1958. Samuelson's role in the conference is fully described in Backhouse 2012. It should be noted, however, that RAND and Cowles were not groups that Samuelson was closely connected with: he neither embraced the mathematical purism of topologists of the day nor was committed to the transdisciplinary community at RAND. Asking Samuelson to link linear programming with economic theory, moreover, shows that for other participants this link was all but obvious. Samuelson, with his J. B. Clark Medal in 1947, had different responsibilities in the economics profession compared to the individuals we treat in this chapter: they were marginalized in the larger economics community in the late 1940s; he was not.

> aggregation problems; the programming models developed by G. B. Dantzig and Marshall K. Wood of the Air Force Department to facilitate the handling of complicated allocation problems under administrative control; the discussions of J. Meade, O. Lange, A. P. Lerner, and others, in the economic literature, of the function of a price system in furthering efficient allocation of resources where decisions are made in many independent units; the discussion of models of technological change by H. A. Simon and others, etc. The unifying element in these diverse problems is the use of models assuming that fixed ratios of inputs and outputs characterize each productive activity—an assumption less restrictive than it seems at first. (PASP, 45, Koopmans)

The theories central to the conference were thus Leontief's input-output model as restated by Samuelson, linear programming and its applications as pioneered by Dantzig's simplex algorithm, and welfare economics as shaped by Bergson, Lange, and Lerner. The list of memoranda that had been circulated prior to the conference accordingly included four papers by Dantzig, Samuelson's and Harlan Smith's comments on Leontief, and several papers by Koopmans on activity analysis. In addition, RAND had prepared and made available to conferees English translations of Schlesinger's "Über die Produktionsgleichungen der ökonomischen Wertlehre" (1933) and Wald's two papers, "Über die eindeutige positive Lösbarkeit der neuen Produktionsgleichungen I" (1934) and "Über die Produktionsgleichungen der ökonomischen Wertlehre (Mitteilung II)" (1935). Other memoranda included Gale's "Convex Cones" and Gale, Kuhn and Tucker, and Brown on computation.

Two years later, twenty-five papers were published in the proceedings of the conference.[3] Part 1, with ten papers, concerned the theory

3 The conference volume *Activity Analysis of Production and Allocation, Proceedings of a Conference* was edited by Koopmans and as a Cowles Commission monograph its cover page noted that it was produced in cooperation with Armen Alchian, George Dantzig, Nicholas Georgescu-Roegen, Paul Samuelson, and Albert W. Tucker. With the addition of new materials, and revisions of many of the papers, the volume was not published until 1951. This permitted the authors to explore deeper connections in a swiftly evolving literature.

of programming and allocation and contained, among other gems, George Dantzig's simplex method for solving linear programming problems and Koopmans's reframing of production theory in terms of activity analysis. Part 2's six papers concerned applications of the analysis, like Marshall Woods's linear analysis of nonlinear growth curves in the aircraft industry. Part 3 presented four papers on the new mathematics of convexity, while part 4's five papers addressed problems of computation of the solutions of programming problems and games.

In exploring these theories, we do so in the context of helping the reader understand the theoretical issues that Arrow, Debreu, and McKenzie tackled in their simultaneous work. Indeed, nearly all of the ingredients of an existence proof were on the conference table, though it was too soon to put the separate pieces together. On the very first page of the introduction to the proceedings to the conference, Koopmans mentioned the early work on existence proofs in Menger's Vienna mathematical colloquium. He referred to Neisser's and von Stackelberg's questions about existence and uniqueness in Cassel's formulation of the Walrasian system (Weintraub 1983). He went on to talk about the papers in Karl Menger's mathematical colloquium, particularly those of Schlesinger and the two papers of Abraham Wald. He then took up the von Neumann paper (1937), which was the first dynamic equilibrium exercise. He noted that "we have dwelt on these discussions in some detail because even among mathematical economists their value seems to have been insufficiently realized" (Koopmans 1951a, 2). Indeed, even at the conference there was no apparent realization that an existence proof in an equilibrium model could enable the analyst to conclude that the problem of interest was solvable. The activity analysis conference was, after all, a conference on applied mathematics rather than a conference on the mathematical integration of economic theory. It was only in retrospect that Robert Dorfman, for example, could say:

> He [Koopmans] was excited by the implications of linear programming for the whole theory of resource allocation, which is the fundamental problem of economics. He perceived clearly that an entire economy could be thought of as solving a vast

> linear programming problem in which the prices that emerged from competitive markets played the same role as the dual variables in Dantzig's theory of linear programming. This implied that the theory of linear programming could serve as a basis for rigorous formulation of the theory of general economic equilibrium. (1984, 294)

Koopmans's activity analysis prepared the ground for the theoretical integration of economic theory via equilibrium analysis, but it did not itself survive that integration. Activity analysis did not enter into economic textbooks. Though it was to be an important help in reconstructing production theory—it was amenable to axiomatization—it did not fit into a general equilibrium framework. It concerned technology and did not take account of *independent* production decisions. In other words, it was linked to optimization theory rather than equilibrium analysis (see Arrow 2008, 165).

Similarly Leontief's input-output models helped create the conditions for equilibrium analysis: Georgescu-Roegen thought of input-output analysis as "the first attempt to apply the general equilibrium theory to the analysis of an economic reality" (Koopmans 1951a, 98). Harvard's Wassily Leontief (1905–99) had created a class of models (input-output) designed for examining interindustry relationships. They were characterized by equations that linked the output of a productive process to a linear equation in the various inputs. The fixed nature of production technology—for example, one widget was produced with *a* units of labor, *b* units of steel, *c* units of plastic, and so forth—contrasted with traditional models in which inputs could be substitutes for one another as their relative prices changed. MIT's Paul Samuelson, a visitor at RAND during spring 1949, began working on a paper on substitutability in open Leontief models, drafts of which were read during the April–May period by both Koopmans and Charles Hitch, his host at RAND. It was only two months before the workshop, on April 11, 1949, that Koopmans decided to include Samuelson's paper with those to be circulated prior to the meeting (PASP, 45, Koopmans). Leontief's model, however, in the years to come would not contribute to the integration of economic theory as general equilibrium theory would do. Even if it was fine empirical applied economics, it was reminiscent of the older empirical regime of Cowles.

Neither would Dantzig's simplex algorithm become a permanent fixture in economics textbooks.[4] Indeed, the conference was the birth of mathematical programming, a collection of ideas and problems that would transform applied mathematics. In later years the conference would be referred to as the 0th conference of the Mathematical Programming Society. Two years later, in 1951, Dantzig would organize what then would be called the first of these conferences jointly with Alex Orden and Leon Goldstein, titled "Linear Inequalities and Programming," at the National Bureau for Standards—a conference where nobody who would become relevant to Cowles or economics was present. The Mathematical Programming Society would not be related to the discipline of economics at all but rather to what came to be tagged as "mathematics and its applications in industry, business, and technology." Its founding members were hardly known among economists. Individuals like A. Orden, J. Abadie, M. L. Balinski, P. Wolfe, and G. Zoutendijk were employed by mathematics departments (see Cottle 2010). Dantzig and Koopmans, the main actors in this conference, would continue working in two different communities; later, when Koopmans received the joint Nobel with Leonid Kantorovich for linear programming, Dantzig would be ignored.

The conference thus opened the gate allowing traffic in two directions: mathematical rigor flowed into economic theory, and economic theory flowed into mathematics reinforcing its "applications." This was made possible by the fact that among the participants there was no clear commitment to the discipline of economics, even as they agreed on a general notion of the economic relevance of their research. What united the conferees was their enthusiasm about the

4 Though Dantzig's algorithm would be cited in textbooks for some time, it would slowly be replaced by other methods. To be sure, for many economists, particularly those who considered economic theory to be a theory of optimal choice rather than a theory of the social structure, linear programming would always represent a "bottom-up" approach to optimization. See, for example, Arrow 2008, 161.

new techniques that had been developed in various institutions inside and outside academia.

## Distancing from RAND

After his time at Princeton Koopmans had remained in touch with Samuel Wilks and Frederick Mosteller (both consultants at RAND), so his own leadership of Cowles was an opportunity for RAND to keep a foot in the academic door of "applied science." The Cowles Commission, as we have seen, had become a hybrid institution somewhere between a university department and a national laboratory. Negotiating the tension of those two different institutions of science proved difficult, as Cowles's intellectual integrity required distance from both. Separating its interests from those of the University of Chicago economists was relatively easy. But Cowles needed to establish distance from RAND as well. The Cowles research program, including the "theory of resource allocation," cannot simply be identified with RAND's vision of subjecting politics and warfare to the principles of rational agency as Abella (2008), Amadae (2003), and Mirowski (2002) have argued. The RAND Corporation, in other words, might have provided resources and financial security to Cowles, but it was of no help to Cowlesmen in confronting the issues of legitimacy they faced within the academic economics community. This insecurity is apparent in Koopmans's introduction to the conference proceedings where he refers to the nature of military funding: "If the apparent prominence of military application at this stage is more than a historical accident, the reasons are sociological rather than logical. It does seem that governmental agencies, for whatever reason, have so far provided a better environment and more sympathetic support for the systematic study, abstract and applied, of principles and methods of allocation of resources than private industry" (Koopmans 1951a, 4).

RAND primarily sponsored research in operations research and game theory. Neither of these two subjects was directly related to traditional economic theory. Thus among the participants there was no clear commitment to the *discipline* of economics, even if they agreed on a general notion of the economic relevance of their research. The enthusiasm about techniques, the unifying element of the conference, could be constructed as an expression of intent to change the discipline of economics only in retrospect. The conference rather was an offspring of various developments in independent disciplines. Next to economics and applied mathematics, the conference would equally be decisive for a new theory in organizational and management science that would come to be taught in business departments about to separate from economics departments during those same years. Arrow, for example, would be the founding member of the Department of Operations Research at Stanford. Debreu, in the economics department of the University of California at Berkeley beginning in 1961, would hardly ever engage with George Dantzig, who launched the Center for Operations Research at Berkeley in 1960 in the *engineering* department. While Arrow always considered economic theory to be a theory of optimal choice, this vision only partially overlapped with Debreu's program of assessing theoretical structures or McKenzie's program of developing and refining them. Arrow would thus include linear programming in economics as a "bottom-up approach to optimization" (Arrow 2008, 161), while for Debreu linear programming was simply a tool to enforce rigor in economics through its role in production theory.

## Distancing from Politics

Activity analysis, which is concerned with organizational questions of production based on systems of equations and programming, suggested that notions of market socialism could be derived from such analyses. Discussing a theory of production in the second half of the 1940s in Chicago unavoidably evoked Lange's theory of planning. It was this historical burden of competitive analysis that increased the pressure of legitimacy on the Cowles research program. From this perspective, Koopmans could not avoid mentioning Ludwig von Mises and Oskar Lange on the first pages of the introduction to the proceedings in the following terms: "Particular use is made of those discussions in welfare economics (opened by a challenge of L. von Mises) that dealt with the possibility of economic calculation in a socialist society. The notion of prices as constituting the information that should circulate between centers of decision to make consistent allocation possible emerged from the discussion by Lange, Lerner and others" (1951a, 3).

When Koopmans circulated the introduction prior to its publication to various participants, Paul Samuelson suggested he simply skip references to Lange and von Mises altogether (PASP, 45, Koopmans). But for Koopmans, the spokesman for Cowles, this question had to be settled rather than silenced. He stressed that one of the main sources of activity analysis was traditional welfare economics which, insofar as it concerned problems of productive efficiency, also concerned planning. For Koopmans planning was not so much a political choice as an organizational necessity. In his understanding two streams of welfare economics began to merge: Lerner's approach in his book *The Economics of Control*, which itself had grown from Lange's work, and welfare economics as it grew out of the problem of social costs treated by Arthur Pigou and Hicks among others: "The underlying idea of the models of allocation constructed is that the comparison of the benefits from alternative uses from each good, where not secured by competitive market situations, can be built into the administrative processes that decide the allocation of that good. This suggestion is relevant, not only to the problems of a socialist economy, but also to the allocation problems of the many sectors of capitalist or mixed economies where competitive markets do not penetrate" (ibid.).

In December 1949 at the Allied Social Science Associations (ASSA) meetings, Koopmans gave a speech to a joint session of the AEA, the American Statistical Society, and the Econometric Society titled "Efficient Allocation of Resources" that echoed his introduction to the proceedings. He reviewed the calculation debate and then referred to the nonideological character of planning by referring to the work of Dantzig and Wood in the context of military planning:

> [T]he earlier discussions had been concerned too much with absolute institutional categories encompassing the entire economy. Even in the capitalistic enterprise economy there are many sectors where the guide-posts of a competitive market are lacking and explicit analysis of the allocation problem is needed.

> Another example may be added to that discussed by Wood and Dantzig. In determining the best pattern of routing of empty railroad cars there are no market quotations placing differential prices on alternative geographic locations of cars. Present arrangements permit this complicated problem to be handled only by administrative direction. (1951b, 457)

Koopmans walked a tightrope over the political fire pit. He urged the use of a theory of planning in ways compatible with a democratic U.S. society. We have evidence of his insecurity as he tried to frame a central notion of activity analysis: prices. On March 1, 1950, Koopmans wrote to Samuelson:

> I have been thinking further about the best terminology for what has been variously called shadow prices, accounting prices, efficiency prices. Of these, I now like efficiency prices best, because it indicates that efficiency is presupposed before the price concept can be constructed. However, the word price still has too much of a market connotation to satisfy me completely. How about the good old word "value"? This, of course, has been abused in various metaphysical senses, and has therefore been avoided for some time by the more careful economists. However, I wonder if it could not by now be re-introduced in what is by now a very proper sense.[5] (PASP, 45, Koopmans)

This insecurity about such a basic question as whether to speak of "prices" or "values" at such a late stage can only be understood with respect to the political poison that planning represented in the McCarthy period.

This confusion regarding terminology would later be at the heart of a debate about scientific priority. Leonid Kantorovich independently had discovered the principles of linear programming in the

---

5 The same caution is present in Koopmans's speech given at the AEA at the end of 1949 that also strongly prefigured his introduction: "Further propositions introduce a price concept which is independent of the notion of a market. The foundations on which this price concept is erected consist only of the technological data (input-output coefficients of all activities) and the requirement of efficiency)" (1951b, 461). "The price concept is found to be a mathematical consequence of an efficient choice of activity levels" (462).

Soviet Union (see Bockman and Bernstein 2008). Whatever might have been the reason for giving Kantorovich and Koopmans a joint Nobel Prize, it produced a debate that echoed the terms of the calculation debate. Kantorovich of course did not use the term "prices," which became a problem once his article was to be translated. Without using this term, could his formulation of linear programming be considered the same discovery? If a theory of socialist planning and a market theory are formally equivalent, this difference of framing would not translate to a difference of credit. Yet Abraham Charnes and William Cooper argued against the equivalent achievement of Kantorovich on the basis of an essentialist notion of prices, assuming that a Stalinist state could not bring forth work worthy of a Nobel Prize. Of course formal equivalence versus irreducibility of an economic theory of socialism and markets had been the core of the calculation debate between Lange and von Mises.

Aside from the concerns raised about the prize, Koopmans had a concern about priority: He considered rejecting the prize since George Dantzig had been neglected altogether. But in the end, instead of rejecting it he donated a third of his prize money anonymously to the International Institute for Applied Systems Analysis in which Dantzig and he were active. For those economists awarding the prize, Dantzig might have not even been considered simply because he took linear programming into noneconomics disciplinary waters.

## The New Beginning

We can now return to our larger theme: how and in what ways the 1949 conference signaled the changes that were to occur in economic theory. First, the conference produced the historical conditions for the future existence proof by taking mathematical values of rigor and axiomatization *for granted*. But second, these values still stood quite apart from economics. As a discipline, economics and the profession's postwar institutions were simply out of focus for the conferees. The conference did not announce the public emergence of a subcommunity with a shared purpose of converting the rest of the profession—not even the Econometric Society let alone the AEA—to their vision of the future of economic research. The conference could be an originating event only because the participants did not quarrel about the possible consequences of their work. They enjoyed living their skills with only implicit agreement on their meaning. They performed as a community despite different institutional backgrounds, different theoretical interests, and different notions of the philosophical justification and pragmatic relevance of their work. Their differences did not require the conferees' attention. In later years they reminisced with pleasure on the conference, recalling their shared commitment to tackle problems inherent in theoretical structures.

Certainly some participants might have believed that their work was going to represent new technical standards in the growing Econometric Society. But they would also have had to know that in 1949 at least half of the Econometric Society, and certainly most members of the AEA, were in no position to appreciate their work. If the late 1940s graduate textbooks and syllabi taught no appreciation of Samuelson's modernization of Keynes and Hicks, how could the conferees believe they could convince economists of the power of the new techniques that went far beyond those of Samuelson and Hicks? Who of the participants would have been willing to proclaim the emergence of the new mathematical economics at an AEA roundtable on mathematics and economics that had been organized at the 1951 meetings with panelists Samuelson, Knight, and Machlup? What was topology to Knight or Machlup? (Samuelson 1952, Machlup 1952).

There certainly were some participants who were willing to *confront* others' views with their vision of social science and its responsibilities. We have already noted that Koopmans acted as the spokesman for the conference, framing the possible interpretations and uses of activity analysis while contextualizing its underlying traditions. As we saw earlier, it was Koopmans who was willing to present his vision in front of a joint meeting of the Econometric Society and the AEA half a year after the conference (Koopmans 1951b). It was he who was active in linking the various emergent theories with the theoretical tradition of welfare economics. It was Koopmans who would take on Cowles's critics at the NBER in "Measurement without Theory" (1947) and who in 1957 would write the only piece that defended the theoretical turn at Cowles in methodological terms, in his famous *Three Essays on the State of Economic Science*. One other piece of work must be mentioned: Kenneth Arrow (1951d), outspoken as he was, wrote but one methodological piece during this period defining and limiting the use of higher mathematics in economics: "Mathematical Models in the Social Sciences."[6]

Only one of the conferees was willing to challenge the emergent Cowles perspective: Oskar Morgenstern. Having been taught economics in Vienna, he had an abiding interest in discussing method, as he previously had done and was known for ever since the 1930s (see Boumans 2012). His conference paper of three pages warned of the lack of accuracy of "observations" underlying the data used in linear programming. Morgenstern, who had never been a researcher at Cowles, some years later made a political issue out of his methodological worries about emphasizing theory at the cost of thinking about data (see also Mirowski 2002, 394ff).[7] There is neither evidence for the belief, nor does the previous contextualization suggest, that other participants would react with strong emotions to Morgenstern's critique. One related point of disagreement might have been Leontief's input-output model and its apparent empirical applications. Noting Leontief's absence at the conference, Backhouse says, "It is . . . impossible not to speculate that the conference had developed in a way that Leontief either did not like or about which he was ambivalent. . . . When compared with the conference proposal and the list of papers discussed, the monograph [*Proceedings*]

6 David Gale wrote a piece, too (1956). The other spokesman of the changes occurring at this stage was Paul Samuelson, not from Cowles. In December 1951, Samuelson gave a paper at the AEA meetings defending the use of mathematics (Samuelson 1952). Two years later, he wrote prominently for the *Review of Economics and Statistics* the introduction to a series of papers on the same issue (among other supporters were Klein, Chipman, Tinbergen, Solow, Dorfman, and, again, Koopmans as the only spokesman for the participants of the activity analysis conference [Samuelson 1954]).

7 In a 1953 letter to Rosson L. Cardwell, executive director of the Econometric Society, about the criteria for being named a Fellow of the Society, Morgenstern wrote: "in my view the Fellow ought to be persons who have done some econometric work in the strictest sense. That is today they must have been in one way or another in actual contact with data they have explored and exploited" (Morgenstern, in Louçã and Terlica 2011, 75). Marschak entered that debate and argued if that be the case, one had to exclude members such as John von Neumann. (Of course this showed how much the Econometric Society had learned to overlook the empirical verve of the later John von Neumann.)

focused on abstract theory at the expense of both statistical data and institutional arrangements" (2012, 32, 38). Given Leontief's absence, however, one might easily imagine that his model was not discussed as a challenge to theory but as an interesting addition to method (see Düppe and Weintraub 2014).

Without controversy about the future of economics, the conference was not seen at the time as a signal event for the larger economics community. Unless economists were regular readers of *Econometrica* they would not have even been aware of these new ideas in mathematical theory, and even statistically interested readers of that journal would have had little connection to the conference's papers. In retrospect, the conference became a marker event because it was both exclusive and informal. The conferees had the freedom to not worry about the historical weight of their work, a liberty that had its source in the presentist problem-directed culture of mathematics departments. RAND's involvement was no warrant of immunity against charges that the conference papers were "not economics."[8]

An illustration of the conference's communal spirit can be found in Koopmans's letter to the authors of articles in the conference volume (July 8, 1949): "Since the subject of the volume has been furthered through conversations and conferences as much as through publications, it is suggested that authors feel free to refer to individuals as the source of ideas whenever there are no publications to which to refer" (PASP, 41, Koopmans). As Emmett has pointed out, this community spirit, among economists, had its own local roots at Cowles: "With Marschak's arrival as research director, a new dimension was introduced: regular biweekly seminars with presentations by both internal and external researchers. . . . Cowles began moving Chicago economics toward social science research as a team project. Members of the commission regularly participated in discussion of research prepared 'in-house'" (2011, 99). The conference adopted this Cowles style, one built into the physical structure of the

---

8 As late as the 1980s H. Gregg Lewis, who had been a Chicago economics department mainstay for almost fifty years, would consistently argue in hiring and promotion cases in his new Duke University economics department that candidate X's work was "not economics." With this distinction he attempted to block the hiring of any candidate from Harvard, Yale, MIT, or Stanford.

RAND office building itself. This new communal intellectual culture was based on shared and uninterrogated standards of techniques. Different and even conflicting visions regarding the meaning of science, and even social science, were to coexist. This is what Hollinger has identified as the emergence of the "scientific community" as the unit of production in American universities (1996, 98ff). This novel regime of truth requires cooperation of specialized intellects with a variety of skills rather than the focused energy of a lonely genius. In an emerging theme of the current narrative, issues of credit and of priority can evolve only in such a world. Insofar as science, in the process of its making, becomes a common effort, a matter of conversation rather than contemplation, the relationship between authors and ideas becomes problematic.

Nicholas Georgescu-Roegen was the first to raise issues of credit and priority with respect to the conference. He claimed that *his* results were "first presented on March 22, 1949 at a meeting of the staff of Harvard Economics Research Project" (PASP, 41, Koopmans, letter to Samuelson, February 3, 1950). That is, Georgescu-Roegen claimed equal credit for Samuelson's theorem on the possibility of substitution in Leontief models. Samuelson replied (ibid., letter to Koopmans, February 5, 1950): "I see no reason to question Georgescu-Roegen's claim to independent discovery of the theorem concerning non-substitutability in the Leontief system. I only wonder that it was not discovered sooner by someone."

This short exchange documents the confrontation of two scientific attitudes. In one there is reference to a world in which individuals have agency in conceiving truth. In the other truth is the outcome of shared projects. In the former individuals claim credit, in the latter it is virtuous to give maximum credit to the larger community. Coauthorship, a production mode that became the common model in the science literature, represents a fusion of these two cultures. It is in the practices of coauthoring that the difficulties of living according to the communal values of science are made manifest, something we describe in the following chapter *en detail* in the work of Arrow, Debreu, and McKenzie. In this new intellectual community in economics the work of these three men would find its home; it was there that issues of credit would emerge in an intellectual culture that was more defined by skills and techniques than by ideas and texts.

The 1949 activity analysis conference created the conditions in which the existence proofs could be written even though conferees had no shared intention of rebuilding economics on new, superior grounds. They lived in a transdisciplinary world of scholars with different national and cultural backgrounds and their careers were dedicated to normal, not revolutionary, science. We described the Cowles Commission, dependent on new government funding via the Office of Naval Research and the RAND Corporation, as a buffer for maintaining the scientific optimism of the mathematics department even as it kept more and more distance from other disciplines, particularly economics. The conference established two important preconditions for the existence proofs to come into being: a shared belief in mathematical rigor (e.g., creating distance from Hicks's *Value and Capital*) and the depolitization of economic theory (e.g., creating distance from Lange's market socialism). Each contributed to the emergent community's shared feeling of liberation, of living in the beginning of a new era. Being part of a community had brought forth a kind of intellectual freedom.

## The Conference's Effects on Arrow, Debreu, and McKenzie

Arrow, Debreu, and McKenzie all passed through Cowles in the 1949–51 period during which time there were continuous discussions about the issues broached at the conference. Despite the diverse career paths of our protagonists, the conference paved a common ground for their nearly simultaneous contributions of 1952–54. In his 1987 interview with George Feiwel, Arrow rather bristled at the remark that if it had not been for him, there would have been no equilibrium existence proof:

> Here was Gérard who had exactly the same thing without me. McKenzie had almost exactly the same thing, not quite as good, but pretty much the same thing. Here we had at least two fine scholars with essentially the same idea and, I think, that if neither of them had existed, someone else would have appeared on the scene. Of course, the von Neumann-Nash tradition had created tools. Once the tools are in place, somebody is bound to pick them up. . . . If I had not done it, somebody else would have . . . and unless the idea was there and ready to

> be expounded and incorporated into the main stream. . . . Incidentally, both of us [Debreu] were influenced by Koopmans's formulation of production (activity analysis), that was another common ground, in this respect for the formulation, not the proofs. As far as consumer theory is concerned, we both started from the same point; it is all Hicks (and Hicks himself is not all that new in this respect). (1987, 195)

Arrow's recollections of that time in the late 1940s and early 1950s suggest that for him the problem of the existence of a competitive equilibrium was "in the air." Once one asserts that a problem is sufficiently well defined and that the tools are "out there and known," Arrow's conclusion follows, namely that anybody could have proved the existence of a competitive equilibrium by 1952–53. The elements were there with the work of von Neumann, particularly his use of the Brouwer fixed-point theorem, Kakutani's 1941 generalization of the main theorem from that found in the 1937 von Neumann paper, and Nash's work that was to appear about the same time as the proceedings of the conference. This was the kind of technical mathematical material that was "in the air" after the activity analysis conference in June 1949.

All three young scholars would later recall how the conference transformed their intellectual lives. Arrow, the only one of our three protagonists present at, and contributing to, the conference, called it a "key step in unifying and diffusing the developments in linear programming and relating them to the theory of general equilibrium." He continued:

> This has been regarded by all those in the field, not only those in the Cowles group, as a decisive event. The exchange of ideas was crucial, as Dantzig has testified in his reminiscences. The papers at the conference called scholars' attention to each other; they clarified the concepts and laid a firm foundation for future work. The first proof of the validity of the simplex method was among its most important products. For the development of general equilibrium theory, the most important paper was Koopmans' in which he developed the theory of production from linear activity analysis. This synthesized all the previous lines of study—fixed coefficients, circular flows, smooth pro-

> duction functions. It was the first time that the relations between resource limitations and the boundedness of the production possibilities set, on the one hand, and between the convexity of that set and the linearity assumptions about individual activities, on the other, were set forth clearly. These two results were crucial in the proofs of existence. (Arrow 1991, 12–13)

For Arrow, Koopmans's activity analysis framed the manner by which production should enter into general equilibrium analysis. The Cowles community's newly shared commitment to convexity analysis (in contrast to differential calculus, the mathematics previously used by economists) was to be the necessary condition not only for the existence proof to be written but for that proof's becoming influential.

Debreu, without having participated in the conference, assessed it in terms similar to Arrow's. As noted earlier, his Rockefeller-funded year-long visit to the Cowles Commission began about four months after the conference. The mathematical work that followed on the conference was congenial to him. The mathematical sophistication of the new ideas, the kind of deep mathematical analysis that underlay much of this work, appealed to Debreu as he recognized that Cowles was his natural intellectual home in America. Years later, Debreu emphasized the shared enthusiasm for convexity analysis that the conference introduced:

> [W]ith the passage of time, that conference has stood out more and more clearly as an important moment in the history of mathematical economics. The theory of production was looked at from new viewpoints; the computation of optimal production programs received emphasis; convex analysis was developed for the needs of production theorists and extensively applied; *the observance of mathematical rigor was taken for granted*; and another demonstration of the fecundity of interaction among the economists, mathematicians, and operations researchers was given (Debreu 1983a, 30, emphasis added).

Cowles liberated many mathematicians interested in the social sciences from the awkward feeling that they had to justify the use of mathematics in economics; instead, its importance could be *taken for granted*. In contrast to his time with Maurice Allais, "at Cowles I came to think, very quickly, that full understanding of a problem required no compromise whatsoever with rigor" (Debreu, in Weintraub 2002, 153).

The conference set McKenzie on a new path as well, even though he had not been a participant. Specifically, in the period just before the conference, some early versions of the conference papers had been circulated and preliminary notices of some papers had appeared. Koopmans's own application of programming to a transportation network had caught McKenzie's eye as it seemed to link some of McKenzie's ideas on patterns of world trade with the new technical possibilities. Reading Koopmans's paper led McKenzie to see that he needed to have training in these new ways of doing theory, and so he applied for a leave of absence from Duke to go to Chicago's economics department as a special graduate student. Thus for Arrow, Debreu, and McKenzie, the conference proved to be a decisive event in their intellectual development.

In this and in the previous chapter we have constructed a context for a traditional discovery narrative. But to solve the scientific puzzle whose pieces were now in view required scientific actors to shape these materials. How and in what circumstances did this happen? How were our protagonists, Arrow, Debreu, and McKenzie, to learn of these new ideas and arrive at roughly the same set of judgments about their applicability to the existence problem? That is the question that will concern us in the following chapter.

Figure 1. Arrow in his early thirties, Stanford.
(Photo courtesy of David Arrow).

Figure 2. Arrow at a Bonn workshop on the "Rolandsbogen," 1990.
(Photo courtesy of Werner Hildenbrand).

Figure 3. Arrow and Debreu, a late encounter. Bonn workshop on the "Rolandsbogen," 1990; from left to right: Peter Schönfeld, Gérard Debreu, Werner Hildenbrand, Kenneth Arrow, an assistant.
(Photo courtesy of Werner Hildenbrand).

Figure 4. Debreu graduating from the Collège de Calais (middle row, second from the left). He recalls Jacques Coignard standing to the left of him; Pierre Debray, middle row at the far right; and Annette Wood, lower row, third on left. It was Jacques Coignard who would give him Allais's book to read and who brought Debreu to the birthday party of his future wife. (Permission Bancroft Library, GDP, 23, 49).

Figure 5. Debreu at the summit. Holidays in 1967,
in Bluche, Switzerland, with a day guide.
(Photo courtesy of Chantal Debreu).

Figure 6. Debreu and Shizuo Kakutani at a workshop in Berkeley, 1974.
(Photo courtesy of Werner Hildenbrand).

Figure 7. The joy of mathematics. Debreu at the university celebrations of his Nobel Prize, October 17, 1983, Evans Hall, Berkeley.
(Permission Bancroft Library, GDP, 23, 17).

Figure 8. Debreu among the Nobelists. From left to right: Subramanyan Chandrasekhar (physics) and his wife, Lalitha Chandrasekhar; William Alfred Fowler (physics); Henry Taube (chemistry) and his wife, Mary Taube; Gérard Debreu and his wife, Françoise Debreu. (Missing Nobelists for 1983: Barbara McClintock [medicine], William Golding [literature], Lech Walesa [peace].) Stockholm, Swedish Academy of Science, December 1983.

(Photo courtesy of Werner Hildenbrand).

Figure 9. Lionel McKenzie and the Duke University Department of Economics and Business Administration, September 1948. First row: Frank de Vyver, Charles Landon, Martin Black, Clark Allen, Ben Lemert. Second row: William Haines, Louis MacMillan, Fred Joerg, Ernest Walker, Henry Lehmann, John Shields. Third row: Carl Clamp, Don Humphrey, Benjamin Ratchford, Calvin Bryce Hoover, Frank Hanna, M. J. Williams. Fourth row: Joseph Spengler, Lionel McKenzie, Edward Simmons, William Ross, Arthur Ashbrook, Lloyd Saville (not present: Robert Smith, Herbert Von Beckerath).
(With permission from the University Archives Photo Collection, box 57, David M. Rubenstein Rare Book and Manuscript Library, Duke University).

Figure 10. Lionel McKenzie and his Rochester economics department, spring 1962. Top row, from left: Ronald Jones, Robert Fogel, Norman Kaplan, Robert France, Rudolph Penner. Bottom row, from left: S. C. Tsiang, Hugh Rose, Richard Rossett, Edward Zabel, William Dunkman, Lionel McKenzie. (Photo courtesy of Economics Department, University of Rochester).

Figure 11. McKenzie drinking from his Duke University mug, Rochester economics department coffee lounge.
(Photo courtesy of Ron Jones).

# PART III
## CREDIT

26. In order to see how something can be called an "existence proof," though it does not permit a construction of what exists, think of the different meanings of the word "where" (for example the topological and the metrical). For it is not merely that the existence-proof can leave the place of the "existent" undetermined: there need not be any question of such a place. That is to say: when the proved proposition runs: "there is a number for which . . ." then it need not make sense to ask "and which number is it?" or to say "and this number is . . ."

**WITTGENSTEIN 1978, 284**

# CHAPTER 6

## THREE PROOFS

In his presidential address to the Econometric Society (given in Ottowa and Vienna in 1977), a revision of which was published in 1981, McKenzie said he wished to

> discuss the present status of a classical theory on existence of competitive equilibrium that was proved in various guises in the 1950s by Arrow and Debreu, Debreu, Gale, Kuhn, McKenzie, and Nikaido. The earliest papers were those of Arrow and Debreu, and McKenzie, both of which were presented to the Econometric Society at its Chicago meeting in December, 1952. They were written independently. The paper of Nikaido was also written independently of the other papers but delayed in publication. (1981, 819)

Besides the Keio narrative, this is the only other historical note that McKenzie appears to have published concerning the 1954 existence of equilibrium paper. These remarks, in Keio and Ottawa, although they tracked the narratives he had shared with his students and colleagues at Rochester, were an historical actor's public reconstruction. In fact there was much more to the story, but it was hidden from his view.

The fuller story of how the McKenzie paper, and the Arrow-Debreu paper, actually came to public attention was left to Weintraub, as an historian of that period in economics, to construct and share with the protagonists. That more complete narrative was written by combining the various authors' personal recollections (Weintraub 1983,

1985) with those of other mathematical economists active in the early 1950s. It concatenated and stabilized in print the stories that the actors had, by the early 1980s, settled upon as to "what had happened in the early 1950s." But that 1983 narrative was limited in its evidentiary base by the few available archival documents bearing upon that period. In the late 1990s, the narrative began to take on different characteristics as the Economists Papers Project at the Rubinstein Rare Book and Manuscript Library at Duke University acquired the Kenneth Arrow papers as well as those of Nicholas Georgescu-Roegen, who, as associate editor of *Econometrica* in the early 1950s, was initially responsible for shepherding both the McKenzie and Arrow-Debreu papers through the refereeing and publication process. Those archival materials, available in the 1990s, allowed Weintraub and Gayer (2001) and Weintraub (2002) to reconstruct the referee process for the Arrow-Debreu paper.

Over the past decade the Economists Papers Project archives received several additional sets of related papers. The new availability of the McKenzie and Robert Solow and Leonid Hurwicz papers now permits a reexamination of some issues previously considered settled; those new materials suggest how the two earlier historical reconstructions ignored some important issues concerning both "priority" and "credit." These issues can be further clarified thanks to the availability of the Debreu papers at the Bancroft Library of the University of Berkeley. That archive contains the letters exchanged between Arrow and Debreu as they were writing their article and thus permits a reconstruction of the different roles Arrow and Debreu played in the process of writing their celebrated joint paper.

Coauthored scholarly work is hardly ever produced as one thought out of two minds. But such work often enters into the body of scientific knowledge as the product of one mind. So it is with the joint work of Arrow and Debreu. In the economics canon they are known as having created the Arrow-Debreu model. Given the remarkable fact that Arrow and Debreu received two separate Nobel Prizes for their work, separated by twelve years, questions of credit suggest themselves: What contribution did each make to the final product? Did they play equal roles? Was it a harmonious division of labor, or was there conflict and compromise? Finally, could each accept authorial responsibility for the final result?

These questions are usually left unasked since what Latour (1987) called "science-in-the-making" is not part of the public record. Scientific (particularly mathematical) publications present new results without reporting on the process of arriving at the results. We, however, can answer these questions about the Arrow-Debreu paper because the evolution of the article is almost completely documented in their extensive letter exchange between their first contact in February 1952 and the article's submission in May 1953. Having never previously met, Arrow and Debreu were to meet only once during this period as Arrow was traveling in Europe during most of the time of their joint work. Debreu, meticulous as he was, kept double copies of this letter exchange, including notes on their only meeting.[1] Debreu also kept a personal chronology of his scientific career documenting the order of events in his conceiving the proof (GDP, additional carton 3). In a largely informal fashion, in what follows, we examine the issues they had to resolve, their negotiations of what to include and exclude, and the compromises they made before agreeing on a final version. As Arrow would recall later in his memorial speech for Debreu: "The experience of collaboration mostly by mail (yes, snail-mail) was one of the most exciting periods of my life, as we argued over each detail, produced counter-examples, and reformulated assumptions. I believe that Gérard felt the same" (2005). We will divide this "most exciting period" in two, before and after December 1952. In that month Arrow and Debreu met for the first time in Stanford, and Debreu, orally presenting the paper at the Chicago meeting of the Econometric Society, encountered McKenzie, who was working on a similar proof.

The structure of our narrative is simple. We will begin to develop a history of their proofs by describing their independent paths in early

---

1 All quotes, if not otherwise indicated, are taken from the Debreu papers at the Bancroft library, Carton 10, Folder "Existence of an Equilibrium in a Competitive Economy," and Folder "Competitive Equilibrium." Debreu did not keep copies of every single draft. Comments that he or Arrow may have written there are not documented.

fall 1951. In our story Arrow was the first to leave a "technical report" of his results at the Cowles library by November 1951 before leaving on an eight-month SSRC research trip to study statistical problems in economic planning in Europe (January and February in Rome, March in Montreux, April in London and Paris, and September in Bergen). We will continue by going through the first phase of the collaboration between Arrow and Debreu in spring 1952. McKenzie worked independently. After McKenzie's and Debreu's simultaneous presentations at the Econometric Society meeting in December 1952, McKenzie likely submitted his article in March 1953 to *Econometrica*, while Arrow and Debreu, in a second phase, revised their own paper: Arrow wrote the introduction and the historical note, and the paper was completed by the end of May 1953. It was then submitted to *Econometrica*.

## Arrow's Path

Even during the time he was writing on social choice, Arrow had not stopped thinking about his abandoned dissertation with Wald. At Columbia in fall 1940 he had read Hicks under Hotelling's tutelage.

> Somehow, when reading Hicks, I got the idea that there was a question whether these [competitive equilibrium] solutions exist. I guess I had been exposed to enough mathematics to know that when one has a system of equations, one worries about existence. I may have been thinking about the problem during the War, but it was after the War that I found out that Wald had worked on this problem. I asked him about it and all he said was: "Oh, yes, that is a very, very difficult problem". I thought that if he found it a difficult problem, it was probably nothing for me to touch. But it sort of remained in the back of my mind. (Arrow, in Feiwel 1987, 194)

Hicks had argued that equilibrium existed because the same number of equations and unknowns were the basis of the competitive economic system: each market required that supply be equal to demand, and so for "N" markets there were "N" market equilibrium equations in the "N" unknown prices. Such an argument, first made by Leon Walras, only made sense if the supply and demand curves were linear since the proof intuition was based on a theorem about independent *linear* equations.[2] But there was no good reason to think that supply and demand relationships were linear. Arrow recalled,

> By the time I got to the Cowles Commission in 1947, there seemed to be more awareness about the existence question and of Wald's work. Patinkin was stimulated enough to write to Wald about the importance of the inequalities in the definition of equilibrium. Wald replied that they were essential to the proof, a point which I hadn't understood, so this correspondence made an impression on me. (Arrow to Weintraub, November 19, 1981)

At Stanford in fall 1949, as acting assistant professor of economics and statistics, Arrow returned to the question about the existence of equilibrium in a competitive economy. He approached the problem in stages. He first worked on the social welfare implications of assuming the existence of such an equilibrium. Welfare questions emerged naturally from his work on social choice, in particular the question of how, in a competitive economy with a very large number of participants, the paradoxical outcomes of preference aggregation did not occur. Arrow thus sought to construct an explicit model of a competitive economy in which, *were he to assume that equilibrium existed*, he could demonstrate that it would be efficient. By the 1940s economists mostly agreed that an equilibrium set of prices could be termed "efficient" if the market outcome obtained by using any alternative set of (nonequilibrium) prices would make at least one individual worse off. That particular definition of efficiency is called Pareto efficiency after Vilfredo Pareto, the Italian economist who first explored the idea. Arrow was thus led to a well-formed conjecture that competitive equilibria were Pareto efficient. Although Abba Lerner had earlier provided a geometric proof of the result, known as "The First Fundamental Theorem of Welfare Economics," Arrow's

---

2 High school algebra courses still teach students how to solve for two unknowns in two linear equations, or three unknowns in three linear equations. Solving for the unknown in quadratic equations is harder and there is no high-school-level method for going beyond quadratics to more general nonlinear equations.

proof laid bare the fundamental assumptions necessary to establish the truth of the conjecture (Arrow 1951c). Though this paper was formulated with little mathematics, Arrow also used convexity analysis and in particular the separating hyperplane theorem, a tool whose importance had emerged in the activity analysis conference. He presented his paper in August 1950 at the Second Berkeley Symposium on Mathematical Statistics and Probability. It would be refereed for the Cowles Discussion Paper series by Debreu, who had just arrived in Chicago after Arrow had left for Stanford.[3]

The significant contribution of Arrow's paper involved constructing a model of a competitive economy that was built up from assumptions about individual behavior, where those assumptions represented the preferences of individuals and firms. Previous analyses, like those of Hicks, failed to assume that prices had to be nonnegative in order to rule out absurd equilibria with negative prices; in Hicks's book, for instance, a competitive economy was simply a collection of competitive markets or, equivalently, a collection of supply and demand equations.

With Arrow's model, and the result that Pareto efficiency would necessarily obtain were a competitive equilibrium to exist, the large remaining problem was whether or not such an equilibrium *could* possibly exist.[4]

---

3 During this period Arrow was also involved in many other projects. One worth mentioning is a paper he wrote at the request of the behavioral scientists Daniel Lerner and Harold D. Lasswell, a methodological piece titled "Mathematical Models in the Social Sciences" (1951d). This was exceptional not only because of his young age but also because there was hardly anyone else in the Cowles circle who reached out to a larger community of social scientists. We can assume that after Koopmans's engagement with the NBER institutionalists (1949), such programmatic discussions were not common at Cowles. Later David Gale (1956) would write a similar piece for the *American Scientist* titled "Mathematics and Economic Models." His efforts would not be well-regarded and perhaps even cost him a job: "I have heard some economists," Samuelson wrote about Gale on the occasion of a job application, "regard as a trifle naïve the paper on Mathematical Economics that Gale wrote for the *American Scientist*" (PASP, 45, Koopmans).

4 Mathematicians' interest in this kind of problem is immense. An old joke, for instance, tells of an engineer, a physicist, and a mathematician at a science convention, sharing a room. A fire alarm sounds. The engineer screams and runs out the door. The physicist places his hand on the door and, feeling no heat, screams and runs out of the room. The mathematician walks into the bathroom, turns on the cold water faucet, watches the water flow, mutters, "A solution exists," and goes back to bed.

> The next crucial stages in my development were, on the one hand, Nash's papers on the equilibrium point as a solution concept for games and on the other the development of production theory on the basis of linear programming, by Koopmans, as you surmised. According to my recollection, someone at RAND prepared an English translation of the *Ergebnisse* papers to be used by Samuelson and Solow in their projected book, (sponsored by RAND), which emerged years later in collaboration with Dorfman. I read the translations and somehow derived the conviction that Wald was giving a disguised fixed-point argument (this was after seeing Nash's papers). In the fall of 1951, I thought about this combination of ideas and quickly saw that competitive equilibrium could be described as the equilibrium point of the suitably defined game by adding some artificial players who chose prices and others who chose marginal utilities of income for the individuals. The Koopmans paper then played an essential role showing that convexity and compactness conditions can be assumed with no loss of generality so that the Nash theorem could be applied. (Arrow to Weintraub, November 19, 1981)

Arrow thus remained true to Nash's game theoretical context. He provided a synthesis of the notion of a game and that of competition by adding a "fictitious player" who chooses prices and marginal utilities. This player would later be one point of disagreement between Arrow and Debreu.

Arrow constructed a first version of the proof: "I finally wrote this up as a technical report in October or November 1951. Then I went off to Europe on a Social Science Research Council fellowship." That report, which Debreu would read and discuss some months later, was called "On the Existence of Solutions to the Equations of General Equilibrium under Conditions of Perfect Competition." It would not be a Cowles discussion paper and, unfortunately, is no longer available at the Cowles library, nor does Arrow possess a copy. "When I got to Rome just before or after New Year, I found at my mailing address a letter from Gérard Debreu whom I barely knew. We had never been at the same place at the same time. He wrote that he had been working on the same problem and he sent me his manuscript" (Arrow, in Feiwel 1987, 194). Though they moved in the same circles, Arrow and Debreu had never met, and they would not meet until a first full draft of the paper was written. What led Debreu to work on the existence proof in general equilibrium theory?

## Debreu's Path

In the months before his arrival at Cowles in June 1950, Debreu had visited Uppsala and Oslo, where he met Ragnar Frisch. It was during this period that he wrote a paper linking Pareto efficiency to equilibrium using calculus, as he noted in his list of accomplishments sent to the Rockefeller Foundation (GDP, additional carton 3). The generalization of this paper, using not calculus but convexity tools, was Debreu's "debut" paper at Cowles in the summer of 1950. In August 1950, at the same time that Arrow presented his optimality proof in Berkeley, Debreu presented his proof of the optimality theorem at the Harvard meeting of the Econometric Society; the paper eventually appeared as "The Coefficient of Resource Utilization" (Debreu 1951a). The paper "provided a non-calculus proof of the intrinsic existence of price systems associated with optimal complexes of physical resources—the basic theorem of the new welfare economics. . . . This proof is based on convexity properties" (ibid., 274). In a "historical note" regarding calculus, he dismissed Pareto for not having established the conditions of an optimum "in spite of lengthy developments" and claimed that "gradual improvements [were] brought by Barone, Bergson, Hotelling, Hicks, Lange, Lerner, Allais, Samuelson, and Tintner" (ibid., 282), though none of them could shed the corset of calculus. In Debreu's view their work represented a half a century of economic theory spent in vain attempts to be rigorous.

One of Debreu's first tasks at Cowles was to review Arrow's optimality proof that, like his, used convexity arguments but was less formal than his own treatment. Debreu recalled: "The Cowles Commission had an internal refereeing process and it is in this connection that I was shown the manuscript of K. J. Arrow's paper by William B. Simpson, the Assistant Director of Research of the Cowles Commission. As I recall, W. B. Simpson asked me whether Arrow's contribution should be included in the Cowles Commission Reprint Series, and also to comment on the substance of the paper" (Debreu, personal communication, in Weintraub 1985, 95). Debreu acknowledged Arrow's article in the published version of the Harvard paper with a footnote pointing out that Arrow, too, had presented a paper that also "contains a non-calculus proof of the basic theorem. Unfortunately I had his manuscript in my hands for too short a time to appraise it fully here" (Weintraub 1985, 95). Arrow in his article (see ibid., 96–97) thanked Debreu for his comments (1951c, 507).

After his arrival at Cowles, Debreu began reading the same articles that used the fixed-point theorem in economics that McKenzie and Arrow were reading. His interests, however, were different. From his Bourbakian perspective Debreu was not much interested in the applied contexts in which fixed-point theorems could be employed in thinking about games, strategies, competition, price-taking, and so forth. He was instead rather taken with the fact that fixed-point theorems were topological since topological structures were one of the "mother-structures" of Bourbaki mathematics. His first encounter with a fixed-point theorem in an economic context might have been von Neumann's and Morgenstern's reference to Kakutani (1944, 154). But his attention to it was certainly fostered by his office mate at Cowles, Morton Slater, who used Kakutani's theorem in his own set of notes (1950) that was read by both McKenzie and Debreu, each of them unaware of the other's interest though they would have seen one another every day. Slater quoted von Neumann's original article (1937) that Debreu must have read around the same time as he read Kakutani. The English translation of Wald's work (1951), prepared at RAND for the 1949 Cowles conference, was readily available to members of the Cowles community in 1950. But even if Debreu read Wald at this stage, it made no impression on him.

> According to my recollection, it was when the [Koopmans] monograph was published [in 1951] that I learned of the existence of A. Wald's papers on general economic equilibrium and only when the English translation of the most important of those papers appeared in *Econometrica*, October 1951, did I get acquainted with its contents. At that time, in the fall of 1951,

> I was already at work on the problem of existence of a general economic equilibrium, and insofar as I can trust my memory of events that took place 30 years ago, the research I did on that question was not stimulated at its inception by Wald's articles, nor, I believe, was it influenced in its development by them. (Debreu to Weintraub, December 7, 1981)

Debreu correctly claimed that his interest and first attempts at an existence proof were independent of Wald's work. Indeed Wald was solving an economics problem. It was von Neumann's paper, as used by Slater, that would have stimulated the Bourbakist Debreu: "The paper by Wald that gave the first proof of existence in the early 1930s did not happen to be important for me. The work of von Neumann on growth turned out to be much more significant since, in particular, it led to Kakutani's theorem" (Debreu, in Feiwel 1987, 249).

As we pointed out earlier, von Neumann's 1937 article was the only time he used a general equilibrium framework. But it was not that which caught Debreu's interest. What appealed to Debreu was that the paper led to Kakutani's generalization, a theorem with no economic context. Besides finding in it implicit approval for his own ideas, Debreu was proud of having found, if not an error, a mathematical slip in von Neumann's paper: his theorem was unnecessarily restrictive for his purpose. "He did not need that powerful tool to prove the theorem that he was after. The separation theorem for convex sets was quite sufficient" (GDP, 4, Leonard). Though this discovery must have been very satisfying for Debreu, he later would comment: "Thus the main mathematical tool for the proof of existence of a GE owes its origin to an accident" (1998, 3). By implication, Debreu never thought about von Neumann's aversion to neoclassical price theory. Nor was he taken aback by reading Nash: "[Nash's theorem] was important, in a way, via Kakutani's Theorem, the fixed point theorem, which has always remained one of the major mathematical tools, in my opinion. But [the] Nash equilibrium has never played an important role for me professionally" (GDP, 4, Leonard).

And so Debreu expected that his own contribution to economics would be to work with a yet more general version of the fixed-point theorem as advanced by topologists. Besides Israel Herstein and John Milnor, he consulted the most prominent Bourbakists at Chicago, Saunders Mac Lane and André Weil. They helped him work with a version of the theorem by Deane Montgomery and Samuel Eilenberg (1946). From Montgomery as well as from Jean-Louis Koszul, another Cartan student and second-generation Bourbakist like Eilenberg, Debreu learned of a yet different version of the theorem by Edward Begle (1950). By January 1951 Debreu had produced his own article on a saddlepoint existence theorem without much reference to an economic context (1951b). He noted that his theorem was equally valid for von Neumann's 1928 and 1937 papers, and was also more general than Kakutani, since it replaced convexity by contractibility assumptions (ibid., n. 3). This discussion paper would be the basis for the comments he would later make on Arrow's technical report. Arrow never sought, or reached, that level of abstraction.

Though Arrow and Debreu came to their joint paper by different paths, they had covered a great deal of common ground before they began their work together. Arrow spoke of a simultaneous discovery: "This was essentially an example of two people arriving totally independently at the same solution" (Arrow, in Feiwel 1987, 195). This discovery had certainly been in the air since von Neumann's pioneering article, but what is important for the present purpose is that after Arrow and Debreu had conceived the idea of a fixed-point proof in general equilibrium theory independently, writing the article together would require only modest negotiations. They would not confront jointly their separate ideas of its deeper meaning.

## McKenzie's Path

How did McKenzie develop the idea of proving the existence of equilibrium theorem? After 1954 (and indeed through to the year of his death in 2010) McKenzie would retell his own story of the origin and development of his model of interrelated markets and the existence of equilibrium. However, McKenzie wrote only one autobiographical piece, the paper noted in chapter 2, which he read on the occasion of his honorary degree from Keio University. That short discussion in its specifics reiterated what he had written in 1981 to Weintraub, who was then writing his *Journal of Economic Literature* survey paper (1983). It had become a stable narrative, a story that over the years made the points about his "rebirth" at Cowles, his working out the proof alone at Duke, and securing priority by publishing his result before Arrow and Debreu. His friends and students "knew" his story from multiple tellings, as Ronald Jones recalled (personal communication), but the Keio paper stands alone as McKenzie's public account of what he believed he had accomplished in 1952–54. He wrote:

> A piece of research I completed at Chicago was done in Koopmans's class on activity analysis and was based on Graham's model of international trade that I remembered from Princeton. It was a multi sector analysis of comparative advantage which showed that bilateral comparison of comparative advantage was not sufficient to discover an efficient allocation of world production. This led to my article in the *Review of Economic Studies* (1954) entitled "Specialization and Efficiency in World Production". Koopmans was pleased with this paper and suggested that I stay longer in Chicago, but I felt I should return to Duke, a decision somewhat like my decision not to pursue the Oxford thesis or the earlier decision not to do physics at Oxford. I had become quite interested in mathematics and if I had stayed in Chicago I might well have transferred to the math department (1999, 5).

Two points are worth noting. First, although McKenzie wrote the "Specialization" paper in Chicago and says he completed it in Chicago, he in fact finished the article based on that paper only after he had returned to Duke in June 1951. As he would later say, he finished working on that paper before he began working on the existence of equilibrium paper at Duke in early fall 1951. Second, although the "Specialization" paper was not "part of" his existence theorem paper, it was a remarkable early use of the tools of activity analysis to answer a sophisticated, and well-known, problem in economics. It employed the efficiency arguments that activity analysis had unpacked and relied heavily on a number of the papers in Koopmans's (1951a) record of the conference, as well as Cowles papers that had developed from those proceedings. McKenzie had moved far beyond the nonmathematical exercise of his *Economic Journal* paper on ideal output and his Princeton and Hicksian upbringing.

McKenzie had responded to Koopmans's suggestion that he stay on in Chicago by trying to find out if that were indeed feasible. He wrote to his Duke chair, Calvin Bryce Hoover, and received a reply from Frank de Vyver, acting economics chair at Duke, dated January 12, 1951:

> I hardly know what to say about your letter of January 4 because we want you to do what you want to do, and I want you to understand that you are a member of our department and we are looking forward to having you back. On the other hand, if you felt that it would be wise for you to continue another year at Chicago and if you find it possible to finance the program, we would be able to make arrangements for you to be off on leave another year. . . . I do not share your feeling that you are not unanimously cherished at Duke. Of course, you know the pressure on production that is around us all the time, but I think you are interested in creative writing and I am sure that there will be no objection from anyone here. Rest assured, however, that if you ask for another year's leave that you will in no way prejudice your position at Duke. At the present time I am acting chairman of the department, so that makes this official. Furthermore, [Chairman Calvin Bryce] Hoover always backs me up if I make decisions while he is away. (LWMP, 6)

McKenzie was ambivalent about returning to Duke. De Vyver's response to McKenzie's worry about his "not being unanimously cherished" is an oblique observation about his junior position in a department of antitheorists and their mathematics-phobic colleagues. Comparing his staying in Chicago with his earlier thoughts about remaining in Oxford suggests that he considered those decisions to be similar. He was married with children, needed employment, and changing fields or direction once again might have seemed self-indulgent, a characteristic unbecoming to him. He saw clearly that he would be better-off professionally in economics were he to remain at Chicago, but he was more attracted to the mathematicians and mathematical economists there like Marschak and Koopmans than to the economists in the economics department. His decision to return to his job at Duke was a risky one for he would be betting that even at Duke he could do scientific work of the highest order. He was thirty-two years old. He could not remain a student forever. In the Keio paper, McKenzie continued his story:

> On my return to Duke I did not immediately do as Koopmans had suggested to me and consider the factor price equalization theorem of Samuelson in the context of an activities model. The remarks Morgenstern made on the existence problem at Princeton and those made by Koopmans at Chicago had interested me in that question. I knew that Graham had given his model of trade to von Neumann to ask for a way of solving for the equilibrium and von Neumann had replied that no analytic solution was possible. I found the Wald and von Neumann papers from the Karl Menger Seminar in the Duke Math Library[5] and read them with my rather weak German. At this time, I wrote my paper "On Equilibrium and Graham's Model of World Trade and Other Competitive Systems." (1999, 4–5)

McKenzie returned to Duke in midsummer 1951 and began working much harder, and with more success, than he had done prior to his Chicago sojourn. With the "Specialization" paper essentially finished, and with Koopmans's encouragement, he quickly submitted it to the *Review of Economic Studies*.

By early fall 1951 he turned his attention to the problem of establishing the existence of equilibrium in Graham's world trade model. By the following fall 1952 he had his result and submitted a proposal to present his paper at the December 1952 Chicago meeting of the Econometric Society. It was thus in the period fall 1951–fall 1952 that McKenzie, at Duke and working alone, did the background reading of the Wald and von Neumann papers from the Menger seminar

---

5 When Weintraub wrote his 1983 paper, he also went to the Duke Math Library to find the Mengerkreis *Ergebnisse* papers of Wald and von Neumann. At that time in the 1980s the math library still was changing over to computerized record-keeping, so when he was given the volume, it had a card in the back listing previous borrowers' names: the only other borrower had been McKenzie.

and worked with the mimeographed notes on fixed-point theorems by Morton Slater that he had brought back with him from Cowles.

> I knew their [Neumann's and Wald's] papers when I wrote my own existence paper in preparation for the Econometric meetings in 1952 in Chicago. Arrow and Debreu [actually Debreu] presented their paper in the same meeting. My knowledge of Kakutani came from a Cowles Commission Discussion Paper, unpublished, by Morton Slater describing Kakutani's fixed point theorem. I was working on Graham's model from the viewpoint of efficient specializations when the idea of proving existence by a projection of the demand point on the production frontier and passing from that point to the price support occurred to me. I had got some notes on Convex Sets from the seminar of Marston Morse, 1949–50, to help also. They had been advertised on the bulletin board of the Chicago math department in 1950–51. (McKenzie to Morgenstern, November 15, 1976, OMP 52, Neumann)

Debreu was inspired by the same sources. McKenzie, however, was not aware that translations of the two Wald papers had been done at RAND in the summer of 1949 for circulation to a few participants in the June 1949 activity analysis conference. Nor did he know anything about the attack on the existence problem, and related problems, by Arrow and Debreu.

## Arrow and Debreu: February–December 1952

Toward the end of January 1952 Koopmans called Debreu to his office to find out about his current research. Debreu might not have spoken to anyone about his work before. McKenzie recalled, "Actually I did visit the Cowles Commission in Chicago when Debreu was there and working on existence, but he kept this fact secret from me. I asked him what he was working on and he refused to say" (McKenzie to Weintraub, September 8, 2009). No matter how preliminary and incomplete his research was, Debreu could not keep it secret from the research director. He told Koopmans of his work on the existence proof, and Koopmans then gave him Arrow's technical report to read. Debreu was pleased. Some days later, on February 5,

1952, he sent Arrow, then in Rome, detailed comments including an outline of his own approach—neatly typed.[6]

> Dear Dr. Arrow: Koopmans last week handed me your remarkable paper "On the existence of Solutions . . . perfect competition." I have read it thoroughly with great delight and now I take the liberty of sending you a long letter of comments. I hope that the criticism I will occasionally make will not mean, in your eyes, that my admiration for the way in which you overcame the difficulties of this subject is lessened.
>
> I had been working myself intensively on this problem for some time when your paper reached me but I had not yet obtained a complete proof of the existence of equilibrium. After having read your article I easily bridged the last gaps in my work. I will give you, below, a concise account of my line of approach, a little different from yours.

In his comments, Debreu provided meticulous notes for simplifying the proof (e.g., by describing technological possibilities not by convex sets but by convex cones) and on how to avoid certain restricting assumptions (e.g., replacing "convexity" with "contractibility"). More critically, he noted an actual error regarding the discontinuity of the minimum worth condition when prices are zero. But the major criticism Debreu addressed in this first letter was that he did not approve of Arrow's use of the "fictitious player": "The introduction of the fictitious players . . . with the use of Kuhn and Tucker's theorem seems artificial to me (and this is probably my most important criticism). The approach I have taken below gets around this."

By using the fictitious player, Arrow remained true to Nash's own game-theoretic framework for the existence proof. Debreu would not

---

6 During the period of their collaboration, both Arrow and Debreu worked on other projects. In Paris, at the colloquium on econometrics, Arrow presented an article on uncertainty, formulating what came to be known as his contingent commodity approach (1952). He received comments by Debreu's mentor Maurice Allais. Debreu, meanwhile, completed a joint paper with the mathematician Israel Herstein titled "Nonnegative Square Matrices" in February 1952 (1953). Inspired by Herstein's and Milnor's only foray into economics (1953), he worked on preference orderings and completed a paper in April 1952 (1954). He also wrote on optimization, completing a paper in January 1953 (1954). These latter contributions would become important for his *Theory of Value* (1959) but were unconnected with his work with Arrow.

succeed in convincing Arrow to abandon that "applied" framework. The fictitious player would appear in the article but would not appear later in Debreu's *Theory of Value*. It would remain a flash point: the fictitious player, for example, permitted game theorists to interpret the Arrow-Debreu model and theorem as relevant for a socialist economy (Shubik 1977).

> Don Brown called my attention to an early paper written by Gérard and to the fact that Ken and Gérard had first looked at a noncooperative game model for proving existence of a competitive equilibrium. I went back to their papers as I wanted to compare the model of a noncooperative model that I cooked up with Gérard's model. There is no question that he was using the noncooperative theory. Specifically, Debreu introduces a fictitious strategic player called "the market." This player is a price namer, i.e. his strategy is to name a price for every good and he is meant to maximize his "profit." The remaining n players are quantity strategy players. Debreu, of course, gets an immediate competitive equilibrium which is also a non-cooperative equilibrium. If you replicate the game one would run into difficulties replicating the market player. You never need more than one market player. . . . In other words, the model that has been published by Gérard is the perfect model for a centralized Soviet economy being run by a price system. (Shubik to Shapley, March 16, 1972, in GDP, 5, Shubik)

Debreu never responded to interpretations like Shubik's.

A week later, on February 12, Debreu sent Arrow a short note identifying Arrow's error and added: "I will finally write a discussion paper about this question and send it to you as soon as it is ready." This paper was the Cowles Discussion Paper 2032: "An Economic Equilibrium Existence Theorem" (1952a). It put Debreu's saddle-point proof, which employed Begle's fixed-point theorem, in an economic context. But he could hardly present an economic motivation for the article. He opens the text as follows: "Economic theory no longer accepts the once standard implication that if the equilibrium of an economic system can be described by a set of equations whose number matches the number of unknowns, an equilibrium point actually exists. A proof of exacting rigor is now required" (1952a, 1). For Debreu, this was sufficient motivation for the paper. Accordingly he closed the text by showing that his theorem implied the theorems of Arrow, Nash, Kakutani, von Neumann, and von Neumann and Morgenstern (ibid., 13). This generalization is what Debreu believed to be his contribution.

Debreu's comments and discussion paper were a challenge for Arrow, to say the least. Arrow was not acquainted with the mathematics of fixed points beyond Kakutani and did not have the privilege of getting extra lessons from André Weil or Saunders Mac Lane. He replied on March 8, 1952, from Montreaux—in barely legible handwriting: "Dear Debreu. I wish to thank you for your series of letters and the manuscript of your Cowles Commission Discussion paper. I am sorry not to have answered earlier, but you can readily understand that work and travel, with the many experiences of a world new to us, have taken my time. Your major point, that my handling of the function $V_i(p)$ is an error, is entirely correct."

Having admitted his error, Arrow found a very similar mistake in Debreu's paper regarding corner solutions: initial endowments need to be strictly positive (1952a, 7, line 10). Debreu immediately added an erratum on March 14. Arrow, however, played down Debreu's mistake: "The defect is very trivial from an economic point of view, since assuming the existence of labor variables amounts to saying no more than that an individual will work if he has no other source of income." Arrow continued by suggesting a new version of his theorem (his "Lemma 2") without accepting Debreu's more general version of the proof and added: "The Theorem just stated is provable in exactly the same way as my Lemma 2. It is, of course, a special case of your theorem, though it has the advantage of avoiding [the] direct hypothesis [of] the continuity of $A_i$ ($\bar{a}_i$), which may be difficult to verify in given situations."

This remark reveals Arrow's perspective. He insisted on his own less general lemma for the sake of its economic meaning and, at this point, even its greater verifiability. Arrow aimed at making the model "work." Assumptions too strong would amount to a failure. Debreu hardly thought that way: which assumptions are necessary or not is mathematically determined, while their meaning is to be assessed ex-post. If they are weak—good for the economist. If they are strong, the proof is valuable for showing how restrictive the model is. Proving existence does not make a model work but rather serves to appraise the model. "In proving existence," he would say later, "one is not trying to make a statement about the real world, one is trying to evaluate the model" (Debreu, in Feiwel 1987, 243).

Having found an error in Debreu's paper, and insisting on his own version of the theorem, Arrow suggested joining forces: "In view of the essential overlapping of results between the two of us, I would propose that we prepare a joint publication. The relation between the two approaches needs some clarification. . . . There should be a still more general function covering both cases, but it may not be worthwhile to investigate. Of course, if you prefer separate publication, it will be perfectly acceptable to me." How must Debreu have felt? At first, certainly a little surprised, since the difference between his approach (via Begle) and Arrow's (via Nash) apparently seemed rather insignificant to Arrow. He also must have felt honored for Arrow had already established a name in the Cowles community. Since everyone around him had greater experience in economic reasoning than himself, he was advised to seize the opportunity to upgrade his economic profile. But his excitement must have been tempered by the fact that his more general proof was not acceptable to Arrow. Arrow made clear that he would not share what Debreu considered his genuine achievement: generality.

Debreu accepted the offer on March 14, 1952, but not without drawing a clear line between what he considered his own mathematical achievement and what he would expect to contribute to Arrow's paper for an audience of economists: he would serve as the article's mathematical engineer. Hence he asked Arrow to agree that he could seek separate publication of his theorem in a mathematical journal:

> The prospect of working in close collaboration with you on this question is very attractive to me and I thank you for your spontaneous offer to write a joint paper. The first point to settle then is the proposal that Koopmans made in his letter of March 11 (of which I am enclosing a copy) of my publishing a

> synthesis of my saddle point paper and of Section 1 of [Cowles Discussion Paper] Economics 2032 in a mathematical journal. There seems to be a definite advantage in excluding the heavier than usual mathematical content from an article written for economists in an economic journal. Secondly, Tucker wrote to us that the replacement of convexity by contractibility in this kind of question was enough of a straight contribution to mathematics to justify publication in a mathematical journal.

Debreu thus justified his own publication by arguing for a separation of the mathematics from their joint paper on "economics proper." From Debreu's point of view, this separate publication needed such a justification since he expected that his contribution to the joint article with Arrow would be close to what he would accomplish in his own publication. From Arrow's point of view, however, there was no conflict whatsoever since Debreu would hardly use the same economics framework. On March 21, by then in London, Arrow replied with a lengthy handwritten letter: "[M]y own efforts in this direction, as given in lemma 2, and the theorem I gave in my last letter, are so much more restricted in scope than your very important contribution, that there can be no hesitation on your part in publishing the results in a mathematical journal."

Instead of pondering the generality of Debreu's axioms, Arrow worried most about a problem regarding the *meaning* of the axioms, particularly those on saturation and public goods as discussed by his mentor, Harold Hotelling: "I would prefer, if possible, not to assume the impossibility of saturation in any one commodity. Hotelling's argument that bridges or museums should be free rests on the hypothesis that individuals will become saturated with those commodities and will not demand infinitely large quantities at zero price." Bridges? Museums? Things got even worse for Debreu when in the same letter, Arrow once more suggested another theorem:

> This theorem can be proved exactly as my Lemma 2. From the mathematical point of view, of course, there is no reason not to make use of your theorem, but from that of exposition and appeal to what is at best a very limited audience, there may be some advantage in this course. Convex sets and Kakutani's

> theorem are beginning to be familiar, and a paper such as ours may accelerate the process, but to appeal to still another fixed-point theorem of still greater generality may not serve a useful pedagogical purpose. We should, of course, refer to your more general theorem.

Arrow thus defended his less general version not only for the reason of greater verifiability but also for pedagogical and rhetorical reasons that were alien to Debreu's mathematical values.

Debreu immediately replied to Arrow, now in Paris, on April 2, 1952. He did not engage with Arrow's worries. Regarding the unsatisfying assumption of strictly positive endowments, he simply referred to authority: "I have naturally taken consolation . . . in the fact that von Neumann has to make also a rather restrictive assumption." Instead Debreu made what was either a spiteful or strategic suggestion: he trumped Arrow's proposal to keep the mathematics to a minimum by proposing to skip the mathematical proof altogether:

> I suggest that in our economic paper we state the preliminary mathematical results with convexity only, that we define naturally all the necessary concepts, but that we give no proof. There will be so much to prove anyhow, and it is certainly highly advisable to keep the mathematical details at their minimum. Are you in general favorable to this?

Although this suggestion would have placed Debreu in the passive role of mathematical proofreader, it was consequential from his point of view: if generality does not count and if economic meaning is all that counts, why bother with a proof at all? As we know, Debreu did not convince Arrow to take that path; the proof would be included. Though Arrow did not accept Debreu's proposal, he gave in to using Debreu's more general (than Kakutani's) version of the fixed-point theorem. Georgescu-Roegen, who later was in charge of *Econometrica*'s refereeing process, would urge the authors "to make the proof more elementary and simpler or to present it as elaborated consequences of other well-known theorems" (Georgescu-Roegen, in Weintraub 2011, 211). In so doing, he referred to Kakutani! Though Debreu would later point to the generality of the proof as *the* distinguishing feature of their paper as compared to McKenzie's proof, in his *Theory of Value* he would introduce the fixed-point theorem by referring to Kakutani.

The April 2 letter ended the first round of negotiations. Debreu planned to write a first draft by May 1952 and expected that it could be presented at the September meeting of the Econometric Society in East Lansing, Michigan. But their work was interrupted. Arrow was busy in Europe, and Debreu suffered from a kidney infection that lasted over a month. He then was incapacitated by the summer heat, as he wrote to Arrow on July 14 apologizing for the lack of progress on the paper. With Arrow's agreement Debreu turned his attention to the synthesis of his two discussion papers (1951b, 1952a), the result of which would appear in August as "A Social Equilibrium Existence Theorem" (1952b). This paper is notable in that it presents the equilibrium without referring to competitive or strategic behavior at all. Debreu simply spoke, generally, of interdependence of behavior. In his conclusion, he made clear what his preferred discourse was when dealing with economics: he praised Begle's, Montgomery's, and Eilenberg's generalizations of the fixed-point theorem "as valuable contributions to topology whose origin can be traced directly to economics" (1952b, 892). While for Arrow mathematics was useful for economics, for Debreu economics was useful for mathematics.

Robert Strotz of Northwestern University, the managing editor of *Econometrica*, had heard of Arrow's and Debreu's project from Koopmans and invited them to present their work at the December meeting of the Econometric Society in Chicago. They accepted and had to agree on a presentable draft. Shortly after Arrow returned from Europe, they finally met at Stanford for their first time in person on Wednesday, December 10, 1952. After their extensive correspondence on the technical matters of the theorem and its proof, they must have grasped that they were working on different intellectual projects. But their work was already so far advanced that it might have been too late to debate the deeper meaning of their proof. Indeed, they did not have much to discuss. Arrow recalled: "It was a wonderful experience, he was just so brilliant to work with. One of us would say a single word, and the other would just understand immediately" (Arrow, in Gallagher 2005). No discussion but immediate understanding? Perhaps Arrow and Debreu immediately understood one another because it was difficult to object to Debreu, who did not have a strong position on economic questions. Whatever transpired on that visit, Debreu left several days earlier than he had planned.

### McKenzie: December 1952–May 1953

Debreu presented the paper in Chicago on December 27, 1952. It turned out that someone else was working on the same problem. Two days later Lionel McKenzie presented a fixed-point existence of equilibrium proof in a model of international trade employing Kakutani's theorem (1954). Debreu must have been impressed, since McKenzie deliberately decided to use Kakutani's theorem while referring to that of Eilenberg and Montgomery (ibid., 158). Debreu believed that his joint paper implied McKenzie's result and, consulting Koopmans beforehand, spoke up about this implication in McKenzie's session. But when Debreu and McKenzie were asked later about this, their memories diverge. We have three different accounts of this event and all three differ from Debreu's. The first time McKenzie was asked about these events was in 1982. He would return to the subject both in his 1999 autobiographical paper and in letters to Weintraub in 2009. These accounts vary in the details and in their emotional content. In 1982 McKenzie wrote:

> I recall that Koopmans, Debreu, Beckman, and Chipman were at my session. The Arrow-Debreu paper had been given the [conference] day before and I had stayed away. However, Debreu rose in the discussion period to suggest that their paper implied my result. I replied that no doubt my paper also implied their result. As it happens, we were both wrong. Debreu says he spoke up after asking Koopmans's advice before the session. Later in his office, Debreu gave me a private exposition of their results. (McKenzie to Weintraub, January 6, 1982, p. 3)

In 1999 McKenzie gave a similar account for publication commemorating Keio's celebration of his career:

> Debreu was present at my presentation and made an intervention to suggest that my paper was implied by theirs, which had been delivered earlier in the meeting. Though I had not heard it, I responded that my paper no doubt implied theirs.

> Literally both statements were false. Their paper used consumer utility functions and Debreu's theorem on the existence of a social equilibrium, which depended on the fixed point theorem of Eilenberg and Montgomery, while I used demand functions and the more elementary Kakutani fixed point theorem. I had learned about the Kakutani theorem from a working paper by Morton Slater, the resident mathematician at the Cowles Commission. (McKenzie 1999, 5)

In 2009, however, he again reported the same event in a different way when writing to Weintraub:

> The first thing I knew about his [Debreu's] work was when we both presented papers on existence to the Chicago meeting of the Econometric Society in 1952. I did not hear his paper but he heard mine and alleged there that my result was implied by his. I entered the possibility that his was also implied by mine. Both allegations were wrong, since he assumed that demand functions were derived from continuous preferences and I assumed that they were continuous and at sufficiently low prices would exceed the production limits. *I cited their paper in mine but they did not pay me the same courtesy. Arrow told me that this was because Debreu did not tell him about my paper. Also they assumed free disposal without acknowledgement.* (McKenzie to Weintraub, September 8, 2009, emphasis added)

The differences in the last letter are noteworthy, and we will return to their implications when we discuss matters of credit in chapter 8. However, it is worth pointing out that Debreu's "suggestion" of McKenzie's first letter, which became Debreu's "intervention" in 1999, has become Debreu's "allegation" in 2009. Also, the last three sentences implying that Debreu behaved improperly stand out. Debreu recalled the December Chicago meeting differently:

> I have no recollection of the episode recounted . . . and I cannot testify one way or the other on this matter. I bring this question up because you might have interpreted absence of comment on my part, as an endorsement of the statements that you quote. T. C. Koopmans may possibly remember what hap-

> pened at that session. . . . Another point must also be noted that according to the [Weintraub] account Lionel had not attended the seminar where I spoke and had no knowledge of the Arrow-Debreu paper. It's stated the next day that [he said] his paper implied our result (Debreu letter to Weintraub, March 24, 1982, 2).

Even if Debreu's account is correct—that is, Debreu made *no* intervention in the discussion of McKenzie's paper—there is a troubling feature of the story: Arrow told McKenzie that even during the time his joint paper was being revised in the first half of 1953, Debreu did not inform him about the existence of McKenzie's paper. There is no reason why an intellectually honest person would not tell his coauthor about a paper that was clearly written along similar lines, even if it was not exactly the same, especially if the issue of its being the same result had been discussed in a public forum. There is also the issue, if as McKenzie claims Koopmans advised Debreu on what to say, of why Koopmans did not mention it to Arrow. In whatever way one assesses these accounts, after the Chicago meeting the question of whether the McKenzie and Arrow-Debreu contributions were "equivalent" would move priority issues to the fore (even though "logical equivalence" became intertwined with the notion of "equivalent quality").

After leaving the Chicago meeting, McKenzie reworked the abstract for publication in *Econometrica* and revised the paper itself for publication. He submitted the existence paper to *Econometrica* early in 1953. Perhaps as a result of his submission, he received a letter on March 25, 1953, from managing editor Strotz asking him to referee someone else's paper, something we have no prior record of his having been asked to do:

> Some time ago, a professor H. Nikaido submitted a manuscript to *Econometrica* which was returned to him subsequently for revision. Professor Georgescu-Roegen saw the manuscript during the period of its processing and was stimulated by it to write a follow up note extending Nikaido's work. Georgescu-Roegen then saw the revision of the Nikaido manuscript and has recently submitted his own note. . . . I would be interested

> in your reaction to the Nikaido papers which has not yet been done, [though] my main interest isn't getting your opinion on the quality of the paper by Georgescu-Roegen.[7] (LWMP, 6)

With his existence paper under review in spring 1953, McKenzie continued his extremely productive pace and began preparing another paper to present at the Kingston, Rhode Island, summer meetings of the Econometric Society in September 1953. Robert Solow appears to have been the chair of the program committee, and on April 9 McKenzie wrote to him:

> I have recently extended Wald's theorem to general linear models assuming only continuous demand functions and Kakutani's theorem. I think this has been most significant of extensions since a long period of competitive equilibrium virtually requires linearity. My proof can be neatly illustrated with a two dimensional diagram. It is the general form of the paper I presented on Graham's model last December. (RMSP, 57)

On that developing paper, McKenzie's next message to Strotz on April 25, 1953, mentions in passing that "I've definitely carried Wald's theorem to its ultimate conclusion by including the case of external economies. I hope to have a MS soon." This would be the paper McKenzie presented that September. On May 1, Solow responded to McKenzie on the subject of the paper McKenzie would give at the Kingston meeting:

> I confess to being slightly awe-struck at your rate of production of new theorems, but I am the slow and slothful type my-

---

7 From current perspectives on editorial practices, it was a curious way to run a journal. Nikaido (whom no one in the United States apparently knew or knew of at that time) had submitted a paper. It landed in the hands of Georgescu-Roegen, the *Econometrica* associate editor for "Theory." After handling the paper as an associate editor, passing it on to referees, receiving their reports, and asking Nikaido to revise the paper accordingly, Georgescu-Roegen wrote his own paper responding to, and extending, Nikaido's work. This led the managing editor, Strotz, to ask McKenzie to referee both the original and the revised Nikaido paper but to ignore the Georgescu-Roegen paper. Should one not be uncomfortable having an editor, Georgescu-Roegen, using paper submissions as raw materials for new work to publish himself? "Conflict of interest" seems to fit the circumstances. But "the past is a foreign country. They do things differently there" (Hartley 1954, 1).

self. In any case, the extensions of Wald's theorem that you describe [strike] me as being intensely interesting. By all means get them on record at Kingston. In fact, if you get them written up before then, I would very much appreciate giving me a look at your results. I would have thought that the existence of external economies would enormously complicate the problem. Perhaps it is the assumption of inelastic supply of capital goods that saved the day. If I understand you correctly, the very possibility of saturation with capital might serve the same purpose. This sounds like a great step forward. You apparently know of Debreu's results along this line. Ken Arrow has also been working on this problem, and the two of them are publishing a joint paper to appear in *Econometrica* some time soon. Perhaps the best thing for you to do is simultaneously to plan on reporting your results at Kingston, and on submitting a paper to *Econometrica*. (RMSP, 57)[8]

Despite McKenzie's having mentioned his presentation at the Chicago meeting the previous December, Solow was unaware in May that McKenzie had already submitted that paper to *Econometrica*, months before the Arrow-Debreu submission in the first week of June 1953 (Weintraub and Gayer 2001; Weintraub 2002, 192). It is important to be clear on this matter. Both the Arrow-Debreu and the McKenzie papers established the existence of a competitive equilibrium for suitable general equilibrium models, both papers drew on the Wald tradition, and both papers employed fixed-point arguments. Both papers were presented to a public audience at the Chicago Econometric Society meeting. McKenzie's desire to get something in print quickly to establish his claims for priority on the existence problem appeared to him at the time to depend upon the publication of an *Econometrica* abstract for his Chicago paper. The "Report of the Chicago Meeting, December 27–29, 1952" (in the July 1953 issue [vol. 21, no. 3, 463–90]) recorded that on Saturday afternoon, December 27, in a session titled "Theory of Games" chaired

8 The paper delivered in Kingston would eventually appear in *Econometrica* in January 1959 after McKenzie completed a series of papers that extended his world trade analyses of his two 1954 papers.

by Harold Hotelling, the "Abstract of paper by Arrow and Debreu and of discussion by Savage [was] not available" (473). Thus the Arrow-Debreu paper had neither been published nor abstracted in the open literature by July 1953. That same report, however, recorded, for the "Selected Papers" session of Monday afternoon, December 29, 1953,[9] chaired by Martin Bronfenbrenner, a four-paragraph abstract of McKenzie's paper titled "The Existence and Uniqueness of Equilibrium in Graham's Model of International Trade." The last paragraph of that abstract read:

> From the generality of the proof it is clear that the special nature of Graham's model is irrelevant. The proof actually constitutes a substantial generalization of the results achieved by A. Wald for Cassel's [general equilibrium] model. Also, the proof does not have the rather intricate nature of that used by Wald. The basic source of the added generality and simplicity is the exploitation of the convexity of the set of outputs and the use of a fixed point theorem. (484)

By July 1953 the abstract of McKenzie's paper was published while that of the Arrow-Debreu paper was not, and the McKenzie paper was submitted first, *by several months*, to *Econometrica*. McKenzie must have expected his contribution to be as visible as that of Arrow and Debreu.

## Arrow and Debreu: December 1952–June 1953

How then did Arrow and Debreu arrive at a final draft? After their first and only meeting Debreu sent Arrow extensive comments (undated, see Arrow's letter on January 1, 1953). The first and most important of them reminded Arrow of the separation of the mathematics and the economics involved: "We should make a great effort to make clear the logical structure of the theorems and carefully distinguish *assumptions* and *conditions*. It is probably impossible to succeed completely without excessive pedantry." In later accounts,

9 This is two days (the Monday following the Saturday) after the Arrow-Debreu paper presentation by Debreu, not the next day as McKenzie had remembered. There in fact were no sessions on the Sunday.

Debreu would present this "careful distinction" as the main contribution of their paper (1984). The fact that he, even at this late stage of the collaboration, had to caution Arrow about separating the two contexts shows how alien this separation was to Arrow. Arrow's response to this criticism (January 13, 1954) was: "I don't agree at all. I think, on the contrary, the important thing is to display the interdependence of the mathematics and the economics."

Debreu also called for caution when addressing interpretations, in particular if they are contested. At several points he noted: "Deletion: Controversy about the interpretation of a text." When Arrow, for example, wanted to refer to David McCord Wright, Debreu noted: "Deletions: The main reason is that the fact and the reasonings [*sic*] do not have the character of certainty and sharpness of the rest of the paper. Moreover I think we should keep away from controversy with Wright about Keynes and forced interpretations of ancient texts." Why refer to quarrels regarding interpretations if the axiomatic structure stands without them? Debreu would win that argument. "Ancient texts" like Keynes's would not be mentioned. In the same spirit, Debreu also suggested they not use game-theoretic notions: "I have changed player, strategy to agent, action. It seems desirable to have a terminology different from that of games. Moreover are the words player and even strategy so good?" Metaphorical speech, inviting wrong-headed connotations, ought to be avoided.

Arrow used Debreu's comments in preparing another draft, which was completed by February 5, 1953. They had not yet written the introduction. Keeping in mind the now-known predilections of his coauthor, Arrow began his letter by apologizing that he was unnecessarily wordy:

> Some of my comments on the assumptions were fairly detailed, but I think they are useful in relating the abstract ideas to the raw material of economic reality. I have probably been pedantic in spelling out details of the proof, instead of leaving them to the reader, please make any changes along those lines that you care to. My work was tremendously simplified by the excellent set of notes that you supplied me with, and I want to thank you for them. . . . I have generalized the formulation of

> dividends to permit non-proportional payments. This in no way complicates the proof, and it adds to the realism, since we can treat of preferred stock, bonds, and other forms of corporate financing.

Arrow's concern for realism again makes clear the difference between his socialization as an economist and Debreu's as a mathematician. Arrow, moreover, "knew" the audience for their potential *Econometrica* paper: economists and others who were sympathetic to the use of mathematics in economic theory. This helps us understand his resistance to Debreu's vision:

> The most important deviation is my unrepentant feeling that "excess demand" is a better concept than "net demand." It simplifies expressions any number of times and is basic when dealing with the interpretation of market equilibrium in terms of the law of supply and demand. Why don't you take a little poll among the Cowles Commission people?

The last sentence's jibe might have left Debreu very confused: take a poll? Among Cowles people? What a curious research project for a mathematician to undertake. Arrow clearly considered their work to be a formalization of "the law of supply and demand." There is no reason to believe that Debreu had ever considered this interpretation at all.

Arrow's sense for realism was further expressed in his continued concern about the assumption of strictly positive initial endowments.[10] He noted an objection made by Columbia University's

---

10 Jerry Green, a Ph.D. student of McKenzie's and himself an important economic theorist, expanded upon this point in a letter to the authors:

> The reason for dropping [the positivity] assumption reveals a lot about the intellectual mindset of each of [the] three protagonists, and differs markedly across the three of them. . . . If one does assume strict positivity the proof of existence becomes very simple. . . . All the difficulty [and] all the interest in general equilibrium theory is that while the price space is bounded the quantity space is not. That makes the application of fixed point methods difficult.
>
> I discussed this with McKenzie many times. . . . He believed that many of the commodities in a general equilibrium model do not have a positive price. The set of commodities that are positively priced is endogenously determined. . . . McKenzie was also motivated by his interest in the inter-industry structure of production. He would often say that the intermediate goods in the production process are neither in anyone's endowment nor in final consumption. . . . It would be a major mistake, in his opinion, to develop a theory of general

William Vickrey, who argued that the initial endowment might not be enough to survive: "I suppose one could have equilibrium through non-survival of some consumption unit, but this seems a gruesome solution," Arrow wrote to Debreu with a sense of disappointment. On March 5, 1953, he informed Debreu of his reading of Joan Robinson on this point.

> In regard to Vickrey's objection that the initial holding may not be sufficient for survival, I have run across an interesting passage in Joan Robinson's "pure theory of international trade" in her Collected Economic Papers. In reference to the theory of equilibrium in that field she holds that it may very well not be possible at the existing levels of population. That is, the equilibrating process may operate through the death of part of the population. "The invisible hand works, but it may work by strangulation."[11]

It was such remarks that Debreu had referred to as "forced interpretations."

As it is with most negotiations, actual differences between the parties were brought forward only at the end of the process. Note that until this point, the paper was not embedded in any discursive context. At no time during their work thus far had they discussed their respective views on the context of the proof. On April 13, 1953, Arrow sent Debreu his version of the introduction including the

---

> equilibrium that did not recognize the endogeneity of the choice of productive techniques and with it the recognition that all individual endowments are located on the boundaries of the relevant budget sets. . . . It is this view of boundary-issues that led McKenzie to make assumptions about indecomposability, disposal, and non-satiation—highly non-intuitive assumptions about how the economy as a whole hangs together, and very different from the individualistic assumptions about preferences and production sets of firms.
>
> Arrow, I believe, held similar views. I do not think he was as tied into the ideas about inter-industry production as McKenzie was, but he certainly wanted his theory to accommodate that case. And as a practical matter, to insure the applicability of his model, he did not want to make an assumption like strict positivity that was sure to be false.
>
> Debreu, on the other hand, probably did not care about specific models of production or consumption of different commodities. He was willing to drop any assumption in the interest of greater generality. . . . [The] way the three authors reacted to this most important assumption—seemingly technical but actually economically crucial—reveals a lot about their scientific and personal motivations. (personal communication)

11 Arrow in fact misquoted the sentence from Joan Robinson, which reads, "The hidden hand will always do its work, but it may work by strangulation" (Robinson 1951, 189).

"historical note" (1954, section 6): "If you want to, expand the introduction in any way. I am not too satisfied with it as it stands, but I just ran out of ideas." The article came close to conclusion without Arrow and Debreu ever discussing any of the issues Arrow raised in this introduction and historical remark, that is, the proof's usefulness for "both descriptive and normative economics" (265), the tradition going back to Walras, Cassel, Neisser, Stackelberg, Zeuthen, and so forth, and the relationship between existence and uniqueness of an equilibrium (287ff).

In his reply (April 23), Debreu showed that he had not considered the necessity of an introduction before and, unsurprisingly, proposed a deletion: "I think that a short introduction was quite in order. I have even deleted three lines at top of page 2." These three lines must have included the following: "from the point of view of normative economics the problem of existence of an equilibrium for a competitive system is therefore also basic" (1954, 265ff). But what for Debreu was a mere matter of conciseness was essential for Arrow, who resisted vehemently: "The deletion of the three lines on top of page 2 of the introduction removes the point of the whole paragraph. It was precisely the fact that the necessity and sufficiency of competitive equilibrium of Pareto optimality still left open a loophole in the argument for a price system that led me to study the existence question. I consider the retention of those lines, or at least their meaning, important." Similarly, concerning welfare economics, Arrow insisted twice on references to standard articles like those of Lange (1942) or Hotelling (1938), but Debreu insisted both times on leaving them out and won out over Arrow (see letters April 17, May 4, and May 13). Thus only at the very end did disagreement regarding the economic meaning of their proof become apparent, particularly as it concerned welfare economics.

In the last revisions that Debreu sent some days before submission, on May 13, 1953, he again tried to minimize the mathematics that he would want to claim for himself. Debreu managed to avoid referring to any other prior use of the fixed-point theorem apart from Nash, von Neumann, and his own. Regarding the reference to Eilenberg he argues: "Delete reference to Eilenberg. [H]is paper is too sophisticated for economists in general." Clearly the more mathematics he would manage to exclude from the article with Arrow, the more he could later claim for himself.

Arrow and Debreu 1954 was not the result of a harmonious division of labor between a mathematically inclined economist and an economically inclined mathematician. The making of Arrow and Debreu 1954 was a negotiation between generality and simplicity, on the one hand, at the cost of explanatory and expository efficacy, on the other. Arrow and Debreu, instead of ever arguing that conflict through and confronting their different interests, compromised. Neither of them could really identify with their joint work. Debreu accepted his role as a mathematical proofreader, and Arrow accepted postponing the more urgent issues for later research. As Arrow commented in retrospect: "[T]he final paper is much closer to his than to my version. . . . It is possible that my exposition was a little closer to what economists would understand than what Gérard might have done had I left him to his own devices. . . . I made more of an effort in writing to bring along the mainstream, to explain what the question is, and I was probably the one who suggested the intertemporal interpretation" (Arrow, in Feiwel 1987, 195ff).

## The Referee Process

Arrow's and Debreu's paper was completed on May 20, 1953. They were uncertain whether to send it to Koopmans or to Robert Strotz of *Econometrica* since they knew that Koopmans would not have an internal referee—the Cowles community had already agreed on its worth. In the end they sent it to Koopmans, who forwarded it to Strotz on June 9, 1953. Koopmans explained his not having provided internal referees: "Needless to say that this does not imply any feeling that we should regard this as an over-specialized study. It is addressed to a classical problem in economic theory and brings to it new mathematical tools." In referring to "a classical problem" and "new mathematical tools" in one sentence, this was the first time that Arrow and Debreu became Arrow-Debreu.

For Cowlesmen there was no doubt about the worth of Arrow and Debreu's article. But outside Cowles, there were not many economists able to appreciate the mathematics and not many mathematicians able to appreciate the economics. This predicament led to an actual conflict during the refereeing process, as thoroughly described by Weintraub and Gayer (2001). In a letter from Robert Strotz, then managing editor of *Econometrica*, to Nicholas Georgescu-Roegen dated June 15, 1953, the third paragraph reads:

> I am enclosing three copies of the manuscript submitted by Arrow and Debreu which falls in your department [as associate editor]. I hope you'll be good enough to arrange for the refereeing of this paper and to advise me on it. I should mention that a rather similar paper was submitted some time earlier by Lionel McKenzie and that it has not yet completed [*sic*] processing. As a matter of fact it is being read at present by Leo Hurwicz and John Nash. I suppose, therefore, that those two readers should not be burdened further with the Arrow-Debreu paper. (NGRP, 8)

Georgescu-Roegen chose William Baumol of Princeton University and the mathematician Cecil Phipps of the University of Florida to referee the Arrow-Debreu paper. Phipps had appointed himself mathematical watchdog of economists' models and derivations and bombarded journal editors, and economists like Milton Friedman and Don Patinkin, with minor notes on major papers. He never published in mathematics journals, as his obsession with the meaning of the null set had made him something of a joke in the mathematics community (see Weintraub and Gayer 2001). In fact, Phipps recommended rejecting the paper. In a bizarre aftermath to its actual publication, Phipps insisted that the editors of *Econometrica* publish a letter by him saying that the paper was all wrong. This brought forward a multiparty exchange of letters initiated by Strotz, who attempted to gather a number of responses to the Phipps objection. Among those he asked to comment on the merits of the Phipps objections to Arrow-Debreu were McKenzie and Hukukane Nikaido. These negotiations were not shared with the authors. Arrow and Debreu did not defend their paper against outsiders; indeed they hardly accepted any changes suggested in the refereeing process.[12]

---

12 One of the few remarks that Arrow and Debreu discussed but rejected was Georgescu-Roegen's suggestion of mentioning Leontief. On January 27, 1954, Arrow wrote to Debreu arguing

That process for McKenzie's paper was as curious as it was for Arrow-Debreu. As the date for the Kingston summer meeting approached, McKenzie wrote to Strotz expressing concern about the fate of his *Econometrica* submission. Strotz replied on June 9, 1953, "I am writing to tell you that I have prodded the referees, who seem to be particularly pokey. I do not want to give you any encouragement regarding the speed with which this processing can be brought to a conclusion, but I do hope that reports might start coming in quite soon." It was not to be. Strotz was then to write the embarrassed note of June 23, 1953:

> Dear Mac: I thought I better write to you to explain that despite our recent promptings of the referees of your manuscript, we have to this date heard nothing from them. This is very bad luck. Your paper is in the hands of two different people and neither have [*sic*] so far sent me any word about it. Knowing how busy people are with the conclusion of the academic year, it is not surprising that nothing happened during the month of May or the first part of June; but one would hope that they could busy themselves with it during the past two or three weeks. I wonder what to do in this case: whether to write to them and recall the manuscript or simply to prompt them again and keep hoping to hear. If the paper is to be recalled, this means that its processing must be started once again and the past several months of waiting will be a complete loss. On the other hand, one hates to throw good time after bad. Since you are the one who has the personal interest in the matter, I thought I would write to you to ask your advice. My own recommendation is, I believe, that another prompting letter and a further wait would be in order for I should certainly hope that this would spur them to some immediate action. I am planning to get to Kingston this year; I'll look forward to seeing you there. *It occurs to me that I ought to reassure you in connection with your manuscript that I shall not publish any similar papers submitted after yours was*

---

that Leontief did not fit in since he worked "with a peculiarly simplified consumption structure so that prices become essentially irrelevant; further, Leontief himself did not introduce the non-negativity conditions which are essential. This is really getting into the area of your competence."

> *submitted before publishing yours, provided, of course, that your paper is found to be acceptable.* (LWMP, 6, emphasis added)

Strotz's hopes were soon dashed. He wrote McKenzie on August 6: "I have given up. Letters have gone to both referees requesting the return of your manuscript to this office right away. I hope to God I can have better luck with the next people. I don't know whether this is a matter of concern to you, but let me assure you that it is my intention not to publish the paper by Arrow and Debreu (which has also been submitted) before the publication of your paper (if both are found acceptable). I think this would only be fair to you."[13] Strotz was scrupulous about keeping this promise. The McKenzie paper appeared in the April 1954 (vol. 22, no. 2) issue, while the Arrow-Debreu paper appeared in the July 1954 (vol. 22, no. 3) issue. There is one last bit in this sequence of what must have appeared to McKenzie to be calamitous: Strotz wrote to McKenzie on August 17, 1953, that "as might be expected, a recent demand that your manuscript be returned by the two tardy referees has brought a brief comment from one of them, with the promise of more detailed comments within a few days. I have already arranged for someone else to take over the refereeing of your paper from the other laggard and hope that we can make some speed from here on in" (LWMP, 6).

We need to be clear here about the chronology. McKenzie was concerned about establishing a claim of priority for his work, which he saw to be the equal of Arrow-Debreu's in the specific sense that they each presented their work on the existence of equilibrium problem at the December 1952 Econometric Society meeting. He saw his special claim to be weaker in that he appeared to be modeling not a general competitive equilibrium system but a specific world trade system. But his claim was stronger in its use of the Kakutani fixed-point theorem, which made, according to some readers, far better connection to the underlying economic theory. Moreover, he had specifically pointed out in his *published* abstract of the Chicago paper that the

---

13 Note that this was at a time when physical reproduction of manuscripts was technically limited. No photocopier, no printer, etc. Today Strotz could have simply sent additional copies to other referees while prodding the slowpokes.

restriction to Graham's trade model was irrelevant to the larger issue of the existence of a competitive equilibrium. How then to solidify at least an equal claim to priority? His strategy was to give the expanded paper at the summer meeting of the Econometric Society in Kingston in September 1953. That paper made absolutely clear that its generality extended far beyond the Graham model. The published abstract of the paper, he hoped, would itself appear before either of the published papers by him or Arrow-Debreu. He was thus very disappointed to receive a letter from Strotz on September 29, 1953, which told him the following:

> As for your abstract, you will grieved to know that present plans, which are, however, still tentative, are to cease publishing abstracts of papers given at meetings. A main reason for this is to economize on space in the journal. I believe this will mean that the abstracts of papers given at the Kingston meetings will not appear, although this is not yet quite definite. It was good to see you again [at Kingston] and I look forward to the next time.

The note from Strotz crushed McKenzie's hopes for a double mention prior to Arrow and Debreu. There would be no recognition of his now more powerful theorem in print before both his original paper and the Arrow-Debreu paper appeared.

Finally, on December 14, 1953, eight months after his original submission, McKenzie received the long-awaited letter from Strotz:

> At last I can report to you on your manuscript entitled "On Equilibrium in Graham's Model of World Trade and Other Competitive Systems." The paper is favorably refereed and I am today writing to Professor Frisch [editor of *Econometrica*] to recommend it for publication. I feel quite confident that he will concur in this recommendation. At the same time, it appears that a fairish amount of revision is desirable, although it is thought that the desired revision would not take a great deal of time to effect. What has happened is that your paper has actually been read thoroughly by only two persons. Let me call them referees number one and number two. Number two, in addition to making his own comments, read the comment of

> number one and commented on the comments. I'm enclosing copies of all this material, properly labeled. . . . My goal is to get your paper into the April issue if at all possible. This means that I really ought to get your revision along about the middle of January if this can be done.

On January 18, 1954, approximately nine months after McKenzie's original submission, Strotz enclosed some comments received on that original McKenzie manuscript, and on January 26, Strotz let him know that he had marked up the paper for the printer. He also mentioned that "Frisch has written me about the exposition of the mathematical material in your paper and I am enclosing an excerpt from his letter dealing with this subject with the thought that you, better than I, might take a stab at changing a few things so as to satisfy him." The excerpt that Strotz enclosed from Frisch's letter begins with the sentence "The Lionel McKenzie MS on Equilibrium in Graham's Model on World Trade and Other Competitive Systems is accepted."

There is one final issue to address. The Weintraub and Gayer (2001) paper showed that the original referees for McKenzie's paper were Leo Hurwicz and John Nash. They did not do their jobs. Sometime that summer Georgescu-Roegen, while traveling, suspended his work as associate editor and was temporarily replaced by Robert Solow. Strotz asked Solow to find another referee for the McKenzie paper even as Hurwicz sent in a short positive report.[14] Who was the new referee? That mystery was solved with the discovery (by Beatrice Cherrier) of a remarkable letter of September 13, 1966, from Debreu to Solow:

> Dear Bob: I may have been responsible in 1953 for a misconception which I find to be spreading. I believe I should endeavor to dispel it.
>
> On October 5, of that year, you asked me to referee for *Econometrica* the article by Lionel McKenzie that eventually appeared in the April 1954 issue. Your request put me in an awkward situation, for on June 9, 1953, Kenneth Arrow and I had sent to Robert Strotz for publication in *Econometrica* the joint paper

14 Nash had apparently sent in only a few sentences complaining that McKenzie had not cited his own paper.

> that was published in the July 1954 issue. The results of our joint paper were more general than those of Lionel in several ways and the main mathematical result on which our work was based was also a fixed point theorem for set-valued functions. The difference between the two papers in this respect was that Lionel used Kakutani's theorem whereas we used the theorem I had published in *Proceedings of the National Academy of Sciences*, October 1952 which rests on the generalization of Kakutani's theorem due to Eilenberg and Montgomery. We hoped thereby to be preparing the way for a theorem on the existence of a competitive equilibrium which would not depend on convexity assumptions. My *Proc. Nat. Acad. Sc.* article was sent to John von Neumann on May 29, 1952. The idea of using a fixed point theorem for set-valued functions to obtain an existence proof for the equilibrium of a competitive economy had occurred to Ken and me several months before.
>
> In my referee's report of December 17, 1953, I leaned away from the temptation to tell you all this and tried to evaluate Lionel's paper on its merits denying myself use of the information that I have just imparted to you. As a result, my report was undoubtedly confusing.
>
> I began to wonder whether I should write to you about this matter some seven or eight years ago when I read footnote 1, p. 374, of Dorfman-Samuelson-Solow [*Linear Programming and Economic Analysis* (New York: McGraw Hill, 1958)].[15] I hope I am not too hasty in writing today. (RMSP, 53, Debreu)

The fact that Debreu was the referee selected by Solow to "pick up the pieces" of the McKenzie refereeing mess appears, in retrospect, to be less startling than it would be were it to happen today. The correspondence between Debreu and Solow and the Debreu report itself are entirely fair to McKenzie. The problem, though, is that Solow did not know that Arrow and Debreu were parties to the story. Solow, when recently asked about this matter, pointed out that when

15 That footnote began: "The use of the Kakutani theorem to prove the existence of an equilibrium is McKenzie's idea. See his study of Graham's international-trade model."

> I—in a hurry because of the [refereeing] history—gave Mac's paper to Gérard to referee, I didn't know that he and Ken had submitted a paper on the same subject. Of course, had I known, I would not have done that. Now that I think about it, the question arises: to whom could I have sent the paper? I could have asked Paul, but he was always so busy that one hesitated to burden him. The community interested in and competent in those questions was trivially small. Leo had been used; Nash was a bad choice. Hicks was impossible, as was Allais. The other people you can think of now did not then exist (Herb Scarf, Frank Hahn, Werner Hildenbrand). It really was a tiny coterie. (Solow to Weintraub, May 26, 2010)

We will have more to say about this strange episode, particularly Debreu's responsibility to have recused himself, in chapter 8. For as he began to take an active interest in placing himself in history, he needed to clarify, both to himself and to Solow, his behavior with respect to McKenzie's paper.

## Simultaneous Discovery, Priority, and Credit

Issues of priority and simultaneous discovery intrude throughout our reconstruction of the three proofs of McKenzie, Arrow, and Debreu. The proofs were conceived independently at about the same time in late 1951. It is impossible to say from any published record or unpublished letter or note whether McKenzie or Arrow or Debreu first had had the idea, or had employed the technique, of using a fixed-point theorem to establish equilibrium. The case for "simultaneous discovery" is unassailable at various stages of the production process.

First, our account shows that no priority claim can be established from the initial presentation of the idea to a third party. Arrow, at Stanford, sent his technical report, which apparently had several gaps and errors in its proofs, to Koopmans at Cowles before he left for Europe in December 1951. Koopmans gave it to Debreu. Debreu was talking regularly with Mac Lane and others at Chicago, but in early 1952 gaps and errors remained in his own proof, as confirmed by his letter to Arrow in February 1952. We have no record of McKenzie's showing his own paper to anyone prior to submitting it for presentation at the 1952 Chicago meeting. Consequently no case for priority can be made on the basis of a public examination of the new idea by a third party.

Second, the initial independent public presentation of the existence proofs occurred at the December 1952 Chicago meeting. Debreu twice noted in letters to Weintraub that his paper with Arrow had been presented at a session of those meetings *before* McKenzie's session. But the program committee could have reversed those sessions. Thus, with respect to which paper was first read in public (and of course they were not read in their entirety in the session), they were given simultaneously.

Third, with respect to priority in publication, the referee process for the two 1954 papers was messy. The Arrow-Debreu paper, submitted at least two months after McKenzie's submission, was refereed by William Baumol and Cecil Glenn Phipps, each of whom did his work quickly, with no prodding needed. McKenzie's referees, selected before any were chosen for the Arrow-Debreu paper, were Leonid Hurwicz and John Nash. Whether Hurwicz had then the reputation that he later developed of not reading his mail or replying in a timely fashion is not clear. Nor is it clear that those not immediately connected with John Nash knew of his illness. In any event, McKenzie had bad luck to have drawn these referees.

Strotz's decision to publish McKenzie's paper first, reflecting its submission before the Arrow-Debreu paper, meant that McKenzie could claim publication priority. But the contingencies of the editorial process at *Econometrica* make that claim too weak a reed to support McKenzie's claim to anything except publication priority. Our reconstruction of the chronology of the events based on the newly opened archives compels the conclusion that no priority claim, for either paper, can be sustained. Priority, however, as the following two chapters will show, becomes contested and historically interesting not before but only after other forms of credit are given. Only as the work becomes alive in a community, as the work's influence extends to other works, as citations to the work appear, and so forth, do the events described in this chapter gain significance. As these other forms of credit emerged, our three protagonists would evince different attitudes regarding priority.

# CHAPTER 7
## AFTERMATH

### Purity and Berkeley

Arrow and Debreu traveled on different scholarly paths after publication of their jointly authored paper. For Debreu, the paper with Arrow had delayed further work and extension of what he considered to be his genuine contribution. He sought a more general existence proof. He took a six-month leave from Cowles to return to France, but he continued searching for a different proof and by spring 1954 completed the version that would later appear in his *Theory of Value* (1959).[1] Nevertheless, he was not completely satisfied with it since he considered that proof still too close to what he had done with Arrow. What he was really looking for was an existence proof eschewing the fixed-point theorem altogether. He once again consulted the most prominent Bourbakists in Chicago: Armand Borel, Pierre Samuel, and, of course, André Weil.

> I obtained the lemma in the form in which it appears in *Theory of Value*, pp. 82–83, in the late spring, or early summer, of 1954. The detailed plan of my monograph then became clear to me, and by the end of the summer 1954 the first four chapters

1 Arrow and Debreu did share, at first and on the surface, an interest in uncertainty. Debreu spent the summer and fall of 1953 at Électricité de France in Paris working with his friends Pierre Massé and Edmond Malinvaud: "The theoretical article on contingent commodities that Arrow published in that year (1952) and the applied problems created for Électricité de France by the uncertain amounts of water in hydroelectric plant reservoirs led me to the study of economic uncertainty that was eventually published as the last chapter of my monograph, *Theory of Value*, 1959" (Debreu 1983a). Arrow, however, did not embrace Debreu's approach to uncertainty.

> were completed, and available in [typed?] form. At that time I did not seriously consider publishing my result in the form of an article because 1) I believed that my monograph would be finished in a few months, and would presumably appear in 1955, 2) given the papers that had been written before the summer of 1954 on the problem of existence (in particular [but not only] Arrow-Debreu), the result did not seem particularly deep or original. Be that as it may, I communicated my result to Armand Borel (who was spending the academic year 1954–55 at the University of Chicago in the fall of 1954 in his office in Eckhart Hall), and to André Weil and to P. Samuel (at a lunch at the Weils) in the spring 1955. In both cases, my purpose was to discuss the question whether one could dispense with a fixed point theorem in proving the lemma. (GDP, additional carton 4)[2]

He would never write such a proof. After the Cowles Commission moved to Yale in fall 1955, Debreu completed the most general version he had to offer and sent it to von Neumann for publication in the *Proceedings of the National Academy of Sciences*.[3] But von Neumann was too ill to transmit the article to the editors, so Debreu sent it instead to Marston Morse. That paper, "Market Equilibrium," would appear in 1956. In the intellectual chronology Debreu kept for himself, he was meticulous in noting the order of these events since two other mathematicians wrote very similar proofs during the same period: David Gale (1955) and Hukukane Nikaido (1956). Since Debreu's interest in a more general proof dated back to 1950, it was important for him to notice when *exactly* he conceived of what later came to be known as the Debreu-Gale-Nikaido lemma: this was the second time, after the encounter with McKenzie, that Debreu faced the possibility of losing a claim to the priority of his results.

---

2 This was written later at the request of Robert Aumann, who asked Debreu to give a personal account of the evolution of his proof as presented in his monograph: "Some comments on the history of the Lemma of Debreu-Gale-Nikaido."

3 Authors could not submit papers directly to the *Proceedings of the National Academy of Sciences*. They had to be submitted on the author's behalf by a member of the academy.

David Gale is somewhat the forgotten man in this history.[4] A Swarthmore College graduate in 1943, his 1949 Princeton mathematics Ph.D. was supervised by Albert Tucker. His thesis in game theory was done alongside that of Tucker's other student at the time, John Nash. His interests in game theory and applications led him to accept a position in the department of mathematics at Brown University, where he taught from 1950 until 1965. It was during this period that he authored the important 1960 volume *The Theory of Linear Economic Models*. From 1965 until his death in 2008 he was professor of mathematics and operations research at the University of California at Berkeley and held an appointment as well as professor of economics. As a mathematician one might suppose that he and Debreu would have worked together. However, Gale's focus on applied mathematics was uncongenial to Debreu.

Hukukane Nikaido's contribution raises many complex issues. Nikaido's mentor, Takama Yasui, had been invited by Martin Bronfenbrenner to present a paper at the 1952 Chicago meeting, and he attended both the McKenzie and Debreu sessions (see Weintraub 1987). Working in Japan unconnected to the American mathematical economics community, Nikaido had developed an existence proof of a general competitive equilibrium using the Kakutani theorem, but it was not until he read McKenzie's paper in 1954 that he believed his own paper might find a place in an English-language journal. However, he ran into difficulties. This time it was Arrow who downplayed Nikaido's contribution; he was the referee. On January 12, 1955, he reported to *Econometrica* associate editor Georgescu-Roegen:

> I have just read carefully the paper of Mr. Nikaido. Although it is an excellently written paper, I cannot recommend its publication because of its extremely close overlap with a paper De-

4 Gale was of course not forgotten in the mathematics and applied mathematics community. His work with Lloyd Shapley (1962) created the literature on matching algorithms (e.g., hospital residencies and resident applicants, students and schools, etc.). A member of the National Academy of Science and a recipient of several mathematics prizes, he was elected a fellow of the Econometric Society, the Center for Advanced Study in the Behavioral Sciences, and the American Academy of Arts and Sciences, and held fellowships from both Fulbright and Guggenheim (twice).

> breu and I have published. The technique of proof is almost identical. Such simplifications as exist are due to his having made a stronger hypothesis. It is true that he appeals directly to Kakutani's paper [rather] than as we did indirectly to the more general Eilenberg Montgomery theorem. However, as we note explicitly, it would be quite easy to modify our proof to make use of the Kakutani theorem and we only made use of Debreu's because it is already available in the literature. . . . [It may not be appropriate for me to be involved in this process.] Perhaps it would be better to have some person other than myself and Debreu review the question of publication since it is possible that I am prejudiced. However, in all frankness, I feel quite sure in my position. (NGRP, 8)

Georgescu-Roegen replied to Arrow on January 17: "After a superficial reading, I arrived at exactly the same opinion as yours, and I am glad to have it now supported by someone else. . . . No matter what one can think about the merits of Nikaido's proof, I feel that *Econometrica* cannot afford to devote space to mere analytical refinements" (NGRP, 8). This story ends with a letter from Georgescu-Roegen to Strotz on February 4, 1955:

> Nikaido's proof is somewhat neater and simpler than that of Arrow-Debreu, but I feel that this merit alone does not justify its publication. It would be a very poor allocation of our resources. Indeed, his paper brings nothing new. I understand that this reason may not be well received by Nikaido and that he might feel particularly dissatisfied after he sees a paper dealing only with a new proof of Arrow-Debreu results by McKenzie published in the forthcoming proceedings of the last conference on linear programming. Notwithstanding, I do not see what we can do about it. (ibid.)

Debreu likewise believed that his own proof, completed by the end of 1954 and published in his *Theory of Value*, "brings nothing new." But his book would strongly contribute to his scientific persona, whose work would receive a Nobel Prize. No doubt Nikaido's difficulty in getting his paper published led to his belief that he might have been ill-treated.[5]

In the fall of 1955 the Cowles Commission moved to Yale, leaving behind its quarrels with the economics department at Chicago. Debreu, thirty-four years old, was appointed as an associate professor of economics but without tenure (at Yale, as at Harvard, "professor" was the only faculty rank that carried tenure). His future in the United States was still uncertain. He continued working in mathematics in direct contact with Shizuo Kakutani of Yale's mathematics department (Debreu 1960) and devoted most of his early time there to his monograph (Debreu 1959). Cowles at Yale was politically calm in these immediate post-McCarthy years, and its members no longer faced the ideological hostility of the Chicago economics department. As a result Debreu's purity and disengagement, while it had been important for Cowles in Chicago, was less relevant for maintaining Cowles's intellectual authority. Soon there was a real divergence of interests between Debreu, who advanced mathematical tools whatever the effects on economic theory, and the others at Cowles like its new director, James Tobin, who advanced economic theory with whatever tools were helpful. Debreu's Cowles colleagues expressed greater hopes than did he for the expressive future of mathematics in economics: Herbert Simon began computer simulations, Jacob Marschak moved to information issues with experimental designs, and Koopmans would take up growth theory. With no sign that Debreu would make such connection to economics, the Yale economics department was ambivalent about granting him tenure. Was he or was he not an economist?

With an uncertain future at Yale, Debreu began to explore other positions. Paul Samuelson, though not associated with Cowles, remained at the center of the larger networks of economic theorists and thus had a good sense of Debreu's general reputation in the larger *Econometrica* community. The Berkeley economics department began recruiting Debreu even before his contract with Yale would end, even before his monograph was published. Samuelson

5 The story of this is well told by Aiko Ikeo (2006), based on interviews with both Yasui and Nikaido.

was asked to advise them about his qualifications, responding in a letter dated November 1, 1957:

> Debreu is the perfect French mathematical type: he would no more dream of publishing a conjecture he had not proved than he would dream of walking naked into the classroom. And he would probably prove the theorem in Hilbert or Banach space rather than in two dimensions. His work on determinativeness of equilibrium is beautifully general. (Whether 20 people will be able to appreciate his forthcoming book I sometimes wonder.) (PASP, 45, Koopmans)

After publication in 1959, the *Theory of Value* was read at least by seven people—the reviewers. All of them showed reservations about its purity and noted the regrettable exclusion of monopolies, externalities, and money. Rather remarkably, Debreu's Cowles colleagues who wrote reviews did not restrain their own skepticism about his project. Leonid Hurwicz wrote that the book is "unique in its uncompromising devotion to maintain the clarity and rigor of the axiomatic structure even at the expense of other objectives" (1961, 416). He seemed to think of it as an explanatory theory as he argued that "one's understanding of the problem would have been greatly deepened by examples lacking equilibrium due to the failure of one or another of the assumptions" (ibid.). Martin Shubik, then at IBM but a future Cowlesman, was even less forgiving, expressing an "uncomfortable feeling that it represents a tidying up of old work and problems which will not necessarily provide a stepping-stone for new work" (1961). And although Debreu had rigorously separated the mathematical context from the economics, Shubik concluded in what would become a template for others' critiques of Debreu: "[E]conomics is not mathematics. Rigor is a necessary but not sufficient condition for a valuable contribution to economic theory" (ibid.).

Debreu must have felt misunderstood, but he did not defend himself. The objections were not new to him, nor would he be spared them for the rest of his career. However, the public judgment that he was merely "tidying up old work" appears to have eroded his confidence. In 1958, he had a paper accepted by the *Review of Economic Studies*: "Cardinal Utility for Even-Chance Mixtures of Pairs of Sure Prospects." In writing to Samuelson about the paper he admitted (November 3, 1958) that "what worries me most today is the thought that the problem on which I have been working may have been solved long ago. I would be grateful to you if you could send me all your references to the literature" (PASP, 25, Debreu). Samuelson sent him a list with works from Fisher to Hayek to Georgescu-Roegen, including an equivalent theorem used by Ramsey. Ignoring the economists' contributions to his problem, Debreu replied: "In view of this new evidence I would like to ask you the following question: If you were a referee for the *Review of Economic Studies* (which has accepted my paper two months ago; but I can withdraw it) would you recommend that it be published? If (and only if) your answer is affirmative, I shall proceed with my plan, adding of course Ramsey to my bibliography" (ibid., December 17, 1958). Debreu was uncomfortable in an economist's world. His quandary was that he depended on economists' judgments about economic originality while at the same time he distrusted economists' mathematical sophistication. His Yale tenure seemed to require that the book be an immediate success among economists. The book's reviews were worrisome. At the end of the 1950s, his professional career was in jeopardy: his five-year associate professorship was running out, and Yale had to decide whether to grant him tenure. They decided not to.

Andreas G. Papandreou, economics department chairman at the University of California at Berkeley, had already begun developing a comprehensive program in "mathematical economics and econometrics." The first major appointment he made was that of Roy Radner in 1957, at the time that Papandreou first inquired about Debreu's interest in coming west. With Debreu's need to find a job, Radner convinced his new colleagues to pursue Debreu, and he received the offer in 1960–61 during a year's stay at the Center for Advanced Study in the Behavioral Sciences in Palo Alto. Debreu also had the opportunity to return to France as a "researcher." However, he did not have the "agrégation de l'enseignement supérieur" that was mandatory for a full professorship in economics. Though returning to France would be a step back in his career (his work was not approved of by the Bourbakists, pure mathematicians all), the differences between his own interests and training and the U.S. academic scene were real and troubling. If his Cowles colleagues were unsupportive of his permanency at Yale, his future career in the United States might be unsatisfactory. He thus decided to pass on the Berkeley offer and return to France.

His wife refused to go: "[W]hen we were in France, on the beach, in [1961], he had the possibility to stay in France. He would have taken it if I had said yes. He resented the fact that I used my veto" (Françoise Debreu, personal communication). In May 1961, he informed Koopmans about "his" decision: "If I judge by the time I spent pondering it, it has probably been the most difficult dilemma I ever had to resolve" (GDP, 8). Debreu accepted Papandreou's employment offer and stayed in the United States for the sake of his family, particularly his daughters, who were more adapted to the United States than he was.

Debreu stayed in Berkeley for the next thirty years. There he managed to build up his own research community distinct from, and appreciated by, other economists. Before Debreu's arrival Berkeley was hardly a stronghold for mathematical economics. Indeed, it was no stronghold for any particular school or way of doing economics but remained proudly eclectic. Economic theory was covered by Roy Radner and Dan McFadden. But the other strengths were to be found in Robert Gordon's group doing business cycle theory and labor economics, George Break representing public finance, Dale Jorgensen in econometrics, and Bent Hansen's macroeconomics group. Abba Lerner arrived at Berkeley in 1966 and remained there until his retirement in 1971. That department's openness made it possible for an outlier like Debreu to feel comfortable; he and others could mind and conduct their own business. His discreetness had found its *terroir*. He could claim expertise in his own sphere without inquiring into, conjecturing about, or even challenging the expertise of others. This diffidence precluded his cooperating with John Harsanyi and Ronald Shephard, and with George Dantzig, who did operations research in the engineering department and the business school. But Debreu found a French connection in the mathematics department, Lucien Le Cam, even though its star there in 1960 was Alfred Tarski. David Gale, also at the mathematics department, was interested in "applied" research. Eventually Debreu's most important mathematical connection was to be Steve Smale. A Fields Medalist in 1966 and increasingly interested in economic dynamics through its connection with his research on dynamical systems (1974), Smale proved to be crucial in confirming Debreu's own mathematical prowess. His interest in economics widened the door through which mathematicians could enter economics without having been trained in it.

What had been a coterie in the 1950s evolved during the 1960s into an identifiable school in economics: neo-Walrasianism—although hardly any of Debreu's devotees would have read Walras. Debreu mentored a first generation of mathematicians who moved to economics during the 1960s like Werner Hildenbrand and a second generation who finished their PhDs in mathematical economics during the same period like Andreu Mas-Colell, Hal Varian, and David Schmeidler. His growing community, spread around the globe, was distinct from, and appreciated by, other mathematical economists.[6]

As Keynes had his Monday evening "club," Debreu had his weekly Monday seminar. Apart from regular attendees from Berkeley like Radner, Gale, McFadden, and Thomas Marschak, the spirit was set by his many guests from abroad who were often mathematically better trained than economists in the United States. One of them was Werner Hildenbrand, who was a visiting professor between 1966 and 1971. They became close, first because Hildenbrand's wife was French and quickly befriended Françoise and second because Hildenbrand also knew Bourbaki's book by heart: "It was a friendship between two married couples. The wives got on with each other, and the men could work well with each other" (Werner Hildenbrand, personal communication). A second decisive contact was Jacques Drèze from Belgium, whom Debreu had already known as a visitor at Cowles in 1954. In 1966, Drèze launched the European equivalent of Cowles: the Center for Operations Research and Econometrics (CORE) financed by the Ford Foundation. Drèze tried to recruit Debreu without success; Françoise did not want to live in Belgium. Many of

6 In Denmark there was Karl Vind, in England Frank Hahn, and in New Zealand A. D. Brownlie. An important link was to Robert Aumann in Israel, who built up his group at the mathematics department of Hebrew University. David Schmeidler (Ph.D. 1969) and Bezalel Peleg (Ph.D. 1964) launched their careers there. In the United States, Debreu's first devotees were Hugo Sonnenschein (Ph.D. 1964) and later Andreu Mas-Colell (Ph.D. 1972), whom Debreu would recruit to Berkeley.

those whom Debreu influenced were located at, or at one or another time passed through, CORE: Birgit Grodal, David Schmeidler, Jean-François Mertens, Jean Gabszewicz, Alan Kirman, and Truman Bewley, to name a few. CORE "grew" several other institutions in Europe, and Debreu visited them by the end of the 1960s.[7]

His community also grew through his Ph.D. students. Most of them, apart from Truman Bewley, were not from the United States, and they continued their careers back in their home countries.[8] None of them, however, would become a new "Debreu." They profited from the authority that mathematical economics had gained but did not direct the profession into yet deeper Debreuvian waters. Most members of his community would not limit themselves to mathematical structures. Sharing his Bourbakist values to some extent, they also showed great interest in institutional applications of equilibrium analysis. As the mathematics versus economics *Methodenstreit* of the late 1940s receded in memory, the new generation of young theorists could pursue careers as economic theorists without having to self-identify as mathematical economists: Truman Bewley could work in capital theory, Thierry de Montbrial in international relations, Jean-Pascal Benassy and Jean-Michel Grandmont in macroeconomics and monetary theory, Tatsuro Ichiishi in the theory of the firm, and so forth.

Even after his definitive 1959 monograph, Debreu continued working on existence proofs in general equilibrium theory. He never stopped seeking generality. In 1962, in "New Concepts and Techniques for Equilibrium Analysis," he generalized the state of the art represented by Arrow and Debreu, McKenzie, Gale, Nikaido, Walter

---

7 In Paris, CEPREMAP (Centre pour la recherche économique et ses applications), going back to Pierre Massé, was launched in 1967. There Debreu was in contact with Monique Florenzano, Claude Fourgeaud, and Bernard Cornet. Hildenbrand, back in Bonn since 1979, promoted Debreu's work in cooperation with Wilhelm Krelle. In 1977 the European Doctoral Program in Quantitative Economics was founded in cooperation with the LSE, Bonn, and later the École des hautes études en sciences sociales, and again later with the University of Pompeu Fabra. In 1981 Jean-Jacques Laffont founded GREMAQ (Groupe de recherche en économie mathématique et quantitative) at Toulouse. Other institutions where Debreu's standards were reproduced on the Continent were, for example, Delta in Paris and CentER in Tilburg.

8 Worth mentioning are Thierry de Montbrial (Ph.D. 1971), Jean-Michel Grandmont (1971), Volker Boehm (1972), Jean-Pascal Benassy (1973), Bryce Hool (1974), Tatsuro Ichiishi, Lawrence Blume (1977), and Beth Allen (1978).

Isard, and David Ostroff. Debreu's intellectual heirs also viewed existence proofs as the supreme discipline (Peleg and Yaari 1970; Mas-Colell 1974; Shafer and Sonnenschein 1975). In the late 1960s Debreu examined more carefully one of the central assumptions (axioms) of his monograph's model: the notion of perfect competition.

When economists write and teach about perfect competition (assumed in the Arrow-Debreu-McKenzie model), there are two intuitive meanings of the phrase that are usually brought forward. The first is the idea that each economic agent "takes prices as given" or makes market choices based on exogenously given prices. But a second sense of the phrase is that each agent is too small to affect market prices. Now certainly we can construct market-decision scenarios in which there are, say, two or three agents and each "takes prices as given" so that the two notions are not "equivalent." It was the remarkable Francis Ysidro Edgeworth in 1881 who constructed an argument that, beginning with two agents "trading" two commodities, added two more agents, then two more, then two more, and so on and concluded that "in the limit" as the number of agents increased, the final market price of the one commodity in terms of the other would be precisely that price that, had the agents taken it as given to them initially, would have left them content not to trade their goods in any other ratio. Put another way, the market equilibrium could be established *either* by having the traders take prices as given or by being infinitesimally small with respect to the totality of agents. *The two equilibrium notions were equivalent in the limit as the number of traders increased.* Recall that a competitive equilibrium for Arrow, Debreu, and McKenzie had been characterized as a set of prices for all goods such that, if the agents were to optimize (maximize utility, minimize costs) with respect to those prices, the market result would be precisely those prices. This is the central intuition behind the fixed-point proof of existence of the competitive equilibrium. On the other hand the Edgeworth conjecture, were it true in a "reasonable" model, implied that one could establish the existence of a competitive equilibrium by an argument whereby prices "converged" on the competitive equilibrium in the limit as the number of agents increased. This idea was first presented by Martin Shubik in 1959 in a game-theoretic model that formalized Edgeworth's classic argument.

Debreu, alert to the implications of proofs of existence of equilibrium that eschewed fixed-point theory, began to investigate this approach with Herbert Scarf, who spent a sabbatical at Stanford in 1961. Together they constructed a generalized Edgeworth model that went well beyond Shubik's to establish the equivalence of strategic (trading) behavior and competitive behavior. The equilibrium notion for the strategic model—an n-person cooperative game—was the "core" of the trading game. The competitive equilibrium was the equilibrium for the model of competitive behavior. In such terms, the problem was to establish the equivalence of the core and the competitive equilibrium in models with large numbers of traders (agents). Debreu and Scarf modeled the infinitesimal trader not by a limit argument but by considering an infinite set of traders ab initio. In such a "space" of traders, each would have literally no size at all, as a point on a line. This kind of characterization introduced the mathematics of measure theory (Debreu and Scarf 1963), a collection of ideas and theories that generalized the common notions of length and volume.

This was not the kind of theory that appealed to Arrow. On February 24, 1971, when Arrow had just completed his rather skeptical book-length treatise on equilibrium analysis with Frank Hahn (1971), he wrote to Debreu:

> Speaking for myself, I am less and less persuaded that the measure-theoretic approach to the core is the only satisfactory one. . . . The principal problem is something that has bothered me from the beginning, about the meaning of equilibrium or of the core when there is a continuum [e.g., an uncountably infinite number of] of traders. Speaking naively, if there are an infinity of traders, endowments are infinite, and it is not easy to know what is meant by equating supply and demand. (GDP, 5)

But Debreu knew all along that the meaning of equilibrium is a subject for interpretation, not proof. Any interpretation of a mathematical object in ordinary language allows the possibility of confusion or paradox. For instance, following the publication of the Arrow-Debreu article, Koopmans had objected that the Arrow-Debreu existence proof did not facilitate any kind of empirical analysis because the equilibrium is empirically underdetermined. But Debreu replied, in responding to Phipps's objections (to Georgescu-Roegen on November 8, 1954), that the fact that it is empirically underdetermined was not a weakness of the proof but its actual achievement: "I conceive of AD as exhibiting the general and abstract feature of a market economy. It is natural that the model has several possible interpretations and it is in fact one of its most interesting characteristics. For example, I have shown last year[9] how a proper interpretation of the symbols gives a theory of uncertainty without any change in formalism" (GDP, 10, Existence). Debreu appreciated the "most interesting characteristic" of mathematics in widening the spectrum of meaning of economic theory. He saw strength where others saw weakness, weakness where others saw strength. Debreu was unlike many mathematical economists in his belief that restricting theory to specific empirical content made theory uninteresting.

It was precisely this cast of mind that led him in 1974, twenty years after the publication of his proof with Arrow, to establish with full mathematical rigor that the existence proof had no strong empirical implication. Simplifying results developed at that time by Hugo Sonnenschein (1972) and Rolf Mantel (1974), Debreu established the structural indeterminacy of excess demand functions. Put simply, if "excess demand functions" are the *observable* market demand functions, the kinds of assumptions made in, say, *Theory of Value* on individual choices place no restrictions on what might be observed at the market level: the model economy of the *Theory of Value* tells us hardly anything worth learning about characteristics of the economy. Debreu was hardly surprised at this result, but others have continued to either praise him (Feiwel 1987) or blame him (Ingrao and Israel 1990) for holding the opposite position. It is as if these results, as critical as they were, scarcely entered the consciousness of those economists who, pro or con, began referring to Arrow-Debreu as the benchmark of rigorous research.

Debreu's 1974 article represents the last entry of his intellectual chronology. Without his ever providing an interpretation of his research program, Sonnenschein-Mantel-Debreu completed it. In his

---

9 He refers to an article published in 1960, "Une economique de l'incertain," written during a stay at Électricité de France during the summer of 1953 (see Debreu 1959, 102).

mid-fifties he walked off the economics stage. He became an American citizen in 1975, was granted the civil rank *chevalier de la legion d'honneur* in 1976, became a member of the National Academy of Sciences in 1977, and, more informally, was proud to lead Berkeley's departmental football team against Arrow's at Stanford. He may have been content with having implemented a sense of rigor in economics without feeling responsible for how the profession continued in other directions. There was no longer any question of being an economist or not. Gérard Debreu had made it.

And so his life calmed down while his family grew. Spending time with his young grandchildren "helped me become more human," he said in an interview (GDP, 5). He also indulged in his hobby, astronomy. We noted earlier that in 1943 he had considered going into astrophysics instead of economics. He did not forget that passion: "By 1982 it seemed that time was softening the edges a bit. . . . My father seemed warmer and less formal; he hiked at Point Reyes, played bridge with his grandchildren, and loved to get out his telescope on starry summer nights and for special events like solar and lunar eclipses" (Chantal Debreu 2005). Stargazing must indeed be a closer experience to the aesthetic appeal of Bourbaki mathematics than doing economic theory: a safe distance from the world—elements and sets, stars and clear nights. In the early 1980s, Debreu must have believed his days in economics were coming to an end.

## McKenzie Creates Rochester's Economics

The existence paper was not the only direct result of McKenzie's Chicago period. In those post-Chicago years, McKenzie was accelerating his production of work in economic theory, writing four articles in 1954 and 1955 on international trade. We have noted his paper "Specialization and Efficiency in World Production," which appeared in June 1954, that melded the activity analysis of Koopmans with the Graham model to explore the efficiency of the trade equilibrium. At roughly the same time as he finished that paper, McKenzie wrote "Equality of Factor Prices in World Trade," which analyzed the factor-price equalization problem set out by Samuelson. He presented it in Montreal at the Econometric Society summer meeting in September 1954, thanking Samuelson, Solow, and Strotz for their advice; it appeared in *Econometrica* in July 1955. That paper has the charming footnote, "The *locus classicus* of activity analysis is Chapter 3 by Tjalling C. Koopmans in *Activity Analysi*" (1955a, 239), proclaiming the new world that had been created three years earlier. And finally in October 1955 McKenzie published "Specialization in Production and the Production Possibility Locus" in the *Review of Economic Studies*, a paper that demonstrated "that the basic flaw in the classical treatment of specialization is neither an assumption about factor supplies nor about constant costs, but the neglect of trade in intermediate products. On the other hand, activity analysis proves to be quite as necessary for the collection of sources of output, when production functions are smooth, as for the selection of activities, when production functions are not smooth" (1955b, 56). With these four papers McKenzie established himself as a trade theorist in the community of mathematical economists. In a few years he had gone from despair about his scholarly future to a self-confidence that he could continue to publish in the very top journals.

Still, he was concerned about his position at Duke and how without a Ph.D. he was ever going to be able to find a better situation with like-minded colleagues and graduate students. He was increasingly frustrated with his situation, believing that he was not much appreciated. He had three young children, a well-educated and professional (though unemployed) Jewish wife in an anti-Semitic North Carolina, and a course load of twelve teaching hours a week. He worried about how he could sustain a serious research career.[10] Even though his existence paper was finally published in spring 1954, it must have appeared to him that the Arrow-Debreu paper would make more of an impression on the nascent theory community. That Solow knew in the summer of 1953 about the Arrow-Debreu paper, which had not been published, but did not know of McKenzie's paper, suggests the recognition asymmetry. The glorious promise of his 1939 Rhodes Scholarship had, fifteen years later, come to naught. Apart from John Hicks in England, whose recommendation would be tempered by

10 "I hope you and Marion don't find 6 children twice as exhausting as we find 3, else God Pity You" (McKenzie to Samuelson, October 29, 1954, PASP, 51, McKenzie).

McKenzie's having left Oxford with only a B.Litt., there were few senior scholars whose support he could seek aside from Tjalling Koopmans, Paul Samuelson, and Robert Strotz.

McKenzie certainly feared that were he to remain at Duke, even his important proof would not give him the visibility of either Arrow at Stanford or Debreu at Cowles. He had no thesis advisor promoting his career; indeed he had no Ph.D. thesis. No one was his mentor, obligated by professional courtesy to sing his praises. His Duke colleagues, though some like Spengler were well-intentioned, did not understand his work. He never had a Ph.D. student at Duke, and the graduate classes he taught were unsuccessful. It was at Duke that his theory class was referred to as "Mad Mac's Mystery Hour" since the students often had no idea what he was talking about, nor did he explain the material clearly to that nonmathematical audience. Indeed, in one such class the students, on their own initiative, stayed for an extra hour as one of the students in the class, who had a mathematics background, went through the entire previous hour's class explaining the material (Craufurd Goodwin [graduate class member], personal communication). Oxford had turned out to be, with respect to the growing theory community in economics, an intellectual backwater, but even there he had initially been regarded by Hicks as dilettantish. In a handwritten footnote by Robert Strotz in his January 26, 1954, letter to McKenzie we read:

> To turn to a personal note, you wrote something on a recent letter concerning criticism of your teaching that you think I may have heard. I won't entirely deny having heard something of that general sort, but I will deny paying any attention to it. What I heard from Bill Allen [at Duke] really wasn't criticism, but I have noted that he is often overly eager to jump on the next guy. (He has a terribly competitive spirit.) May I mention in any event that I recently proposed your name as someone we should try getting to join our department (this is a vote of confidence) and the only objection I encountered was "What the hell courses would we give him to teach that he would be interested in?" I confess I was at a loss for a very satisfactory answer. For

> the record, in case it comes up again, what is your rank at Duke, and where do you stand with your thesis? Are you at all in the market? Bob (LWMP, 6, 1952–56)

Although it would take three more years before McKenzie would answer "yes" to Strotz's question, on September 25, 1954, McKenzie wrote to Samuelson:

> I want to ask your advice on a very confidential basis. I hope you won't regard it as an intrusion, but the truth is, outside Duke, you, Hicks, and Koopmans are about the only senior members of the profession on whom I feel it possible for me to call, and Hicks, of course, is in England and, therefore, not very useful here. What I am wondering is whether I have any possibility of getting to a Northern or Eastern University, and whether you can lend me any assistance in doing so. For numerous reasons, some rather impalpable, I am not entirely satisfied with Duke. For one, I was an undergraduate here. I think too that my teaching has been too closely confined to "mathematical economics" (whatever that is) and econometrics, whereas my interest is simply economic theory. Then I don't think a Southern university can offer me a proper opportunity in general. . . . Now, of course, I have a special disability, since I do not have the Ph.D. . . . I should add that the people here have not been unpleasant to me. Joe Spengler in particular has been very nice. On the other hand, our department is terribly top heavy with full professors, some of them incidentally antagonistic to economic theory. (PASP, 49, McKenzie)

Samuelson would not let him down; he would be the most ardent sponsor of his career. Over the next three years he enthusiastically wrote about McKenzie in a series of references for fellowships or jobs: a Social Science Research Council Fellowship in February 1955, a Fulbright Fellowship in 1956, and faculty positions at Oxford, the University of Michigan, Berkeley, and elsewhere. Samuelson played down McKenzie's "disability," as he did to S. Taylor in Berkeley (December 6, 1955):

> I know that John Hicks thinks very highly of McKenzie because he has told me so in person and has several times written to that effect. None the less after McKenzie had applied for the rarely given D. Phil Oxford degree, the committee there decided to return his thesis to him for further revisions. I gather from Hicks that this act itself is quite rare and is something of an honor; but in any case McKenzie was quite naturally disappointed and has never cared to resubmit a revised version of his thesis. . . . In his research McKenzie is a perfectionist and experience suggests that such people are high in integrity rather than being high in adaptability. (PASP, 49, McKenzie)

After failing in several attempt to get secure employment at one of the leading universities, McKenzie spent the spring and summer semesters of 1956 at the Cowles Foundation, which had by then moved to Yale. McKenzie's own research in the late 1950s and early 1960s solidified his standing among modern economic theorists. In line with Debreu's interests at the time, he tried to derive demand curves without utility functions based on preference relations: "Demand Theory without a Utility Index" (1957). He also continued working on existence proofs integrating demand theory that has been considered the major difference between his proof and Arrow's and Debreu's: "Competitive Equilibrium with Dependent Consumer Preferences" (1955c). After he presented this at a professional meeting, Oskar Morgenstern and William Baumol of Princeton suggested that he submit his several outstanding papers as a doctoral dissertation so that they might award him a Princeton Ph.D. in 1956. He accepted this remarkably generous offer.

During his stay at Cowles, McKenzie was approached by Bernard Schilling, a professor of English at the University of Rochester also visiting Yale, who had been asked by Rochester administrators to speak with McKenzie about joining the Rochester faculty. Since Rochester did not have a separate economics department at that time, neither McKenzie nor anyone else in the small world of mathematical economics would ever have paid attention to the school. But Schilling's conversation concerned McKenzie's interest in creating an economics department there out of an economics "concentration" in business administration, as well as the university's interest in having him start a Ph.D. program in economics. The opportunity to create "his" kind of economics department was exciting, and he then opened discussions with the Rochester administration. In November 1956, less than three years after the publication of his proof and nine years after his intellectual "rebirth" in his visiting year at Cowles, he accepted Rochester's offer. The headline in the university's internal newsletter, *Rochester Reviews* (1957, 4–5), was "Rhodes Scholar Appointed Professor of Economics" and noted that "Dr. McKenzie will be Chairman of the new Department of Economics created by action of the College of Arts and Sciences and approved by the Board of Trustees earlier this year. This move divided the former Department of Business and Economics into separate departments." As McKenzie recalled years later in a faculty speech, "I was attracted by the prospect of building a department according to my own ideas, which, in the event, would mean that economic theory would be the major emphasis" (2012, 227). His use of the term "economic theory" was carefully chosen. He stressed the contrast between economic theory and mathematical economics, distancing his position from that of Debreu, who was to title his own collection of papers *Mathematical Economics* (1983a). Rochester became McKenzie's own place to develop a distinct research community in economic theory.

Even before he took up his appointment on September 1, 1957, McKenzie was recruiting faculty for "his" department. Arts and Sciences Dean W. Albert Noyes[11] kept his word, and McKenzie had the resources he needed. From the teaching staff that did not move to the new business administration department, only two faculty members were suited for the McKenzie era, the labor economist Robert France and previous department chair William Dunkman, at the time on leave in Japan on a Fulbright Professorship.[12] McKenzie

11 Noyes was a distinguished chemist, winner of both the Willard Gibbs Medal and the Joseph Priestley Medal for his work in photochemistry. Rochester, home of Eastman Kodak, knew and appreciated the intellectual strength of its department of chemistry.

12 It is likely that Dunkman was the initial contact person linking the Rochester Ph.D. program to graduate student applicants from Japan, a pipeline that would become essential to the program's growth and success.

liked the idea of being in charge of a program and was looking for middle-rank scholars "handling theory of money and international trade or also growth economics" (PASP, 49, McKenzie, November 19, 1956). Quickly tiring of the struggle to compete with more established economics departments in recruiting "senior" scholars, he began bringing up his own people rather than hiring from the second tier of those available after Harvard and Yale had made their "selection": "The desirability of recruiting senior people is beginning to impress me rather less. I don't really feel that I need any internal assistance and I rather like the idea of bringing up the young. . . . The better I get to know Rochester the more delightful I find the place" (PASP, 49, McKenzie).

He asked his MIT contacts about available junior faculty candidates, and Solow and Samuelson suggested Peter Steiner, Ronald Jones, and Albert Hirschman. McKenzie also considered Gary Becker as well as Albert Ando from Japan but finally settled on Ronald Jones, one of Solow's students. As Jones recalled:

> I first met Lionel in January 1957. He was on leave at Michigan before he was going to be at Rochester that Fall. I was in my first job after MIT, teaching at Swarthmore College (where I had been an undergraduate) on a one year appointment and I knew that I had to find a job. I had an offer from Princeton, and then Lionel got in touch with me. He had talked with Paul Samuelson and Bob Solow and knew that I might be on the market. Then I got a surprising two page letter of recommendation from Bob Solow to me (!) about Lionel. He told me he's a "theorist's theorist." That appealed to me. It sounded kind of nice. What attracted me was that I wouldn't have anyone senior to me in my field (the opposite of how new people on the market now feel). I would have access to graduate courses and graduate students in international trade. (Jones, interview by Weintraub, October 3, 2011)

The next two assistant professors he recruited were Richard Rosett from Yale and Edward Zabel from Princeton and RAND, and he began looking for another person on the senior level "in the direction of policy, development, macro-economics, and such like"

(PASP, 51, McKenzie, April 10, 1958). Later appointments were made to Nanda Choudry and Michio Hatanaka in econometrics, S. C. Tsiang for macroeconomics, Rudolph Penner in public finance, and Alexander Eckstein as well as Norman Kaplan as specialists on communist economies, a field then called "comparative economic systems."[13]

Neither labor economics nor economic history was connected to any Cowles initiatives in the 1950s. Nevertheless, McKenzie made two magnificent appointments with Sherwin Rosen in labor and Robert Fogel, on the advice of Simon Kuznets, in economic history. It was Fogel's presence that would attract Stanley Engerman to Rochester in 1965. Fogel would be the only McKenzie hire to receive a Nobel Prize. Wistfully McKenzie would later say that

> [w]hen the Nobel Prize was awarded Fogel had moved to Chicago, so the fact that he had done the work at Rochester was not always made clear. . . . Thus the early department had distinction in economic theory, international economics, and comparative economic systems. Also the basis was laid for achieving distinction in the fields of econometrics, labor economics, and economic history. (2012, 233)

None of these fields apart from economic theory overlapped with the interests of Gérard Debreu at Berkeley.

Noyes also kept his promise to provide resources for a Ph.D. program in economics. With Jones, Zabel, and Rosett, McKenzie launched the graduate program covering the fields of general equilibrium theory, the theory of the firm, econometrics, international trade, and labor economics. The first cohort of students included Akira Takayama from Japan, Emmanuel Drandakis from Greece, William Bennett, and Arnold Saffer. The link with Japan was strengthened via Nikaido at Osaka University and Zabel's friend from Kobe Commercial University, Fumitoshi Mizutani. Other Japanese students followed (Akihiro Amano, Hiroshi Atsumi, Nobuo Minabe, and Yasuo Uekawa). In the 1990s McKenzie remarked to some colleagues at the

13 Later during the 1970s Rochester also became to be known in macroeconomics as the home of Karl Brunner, Robert Barro, and Robert King.

weekly departmental tea that he might have won the Nobel Prize had his Ph.D. students not been mostly from Japan. Indeed after his retirement his office and personal library were moved to Japan and installed in the Institute for Economic Research at Kyoto University.[14]

By the early 1960s McKenzie had built up the Rochester research community with its own identity, distinct from that of Berkeley, and proud to be labeled *theoretical* as opposed to *mathematical*. It was an excellent time to build such a department. With the Soviet *Sputnik* launch in 1957, the American educational system became obsessed with "catching up to the Russians." Federal money to support the new focus on mathematics and science washed over high schools and universities, and bright students were channeled into such technical fields of study. Rochester welcomed mathematics and science "majors" into its graduate program as over the 1960s it became known as an exciting place for them to study economics and political science. As the founder of the department, McKenzie was its dominant figure; he ironically referred to himself as its "benevolent dictator." The Rochester economics community bonded in a self-conscious manner over common activities like bowling, softball, picnics with students and faculty, and annual "theatricals." Such prideful identity had never characterized Berkeley's economists and their students, members of a very large university community who did not share a hard annual winter together.

It was McKenzie's persona that shaped the community. The Wednesday afternoon departmental tea recalled standard practice in mathematics departments, and Oxford. The students he recruited,

---

14 There was another element to McKenzie's growing importance after his arrival at Rochester. The same trustee-administration initiative for the social sciences that had created the graduate program in economics had also determined that political science would be identified as a department that Rochester could nurture. Dean Noyes, with a real scholar's knowledge of intellectual merit, hired William Riker in 1962 on terms similar to McKenzie's to create a major department and graduate program. Riker had been at Palo Alto's Center for Advanced Study in the Behavioral Sciences in 1960–61 applying game theory to decisions in politics and had connected with Arrow during his fellowship year (see Aldrich 2003). In 1963 Riker brought Duncan Black to Rochester as a visiting professor for a year, giving notice that the political science department intended to forge links with the new positive political theory based on game theory and the theory of social choice (Riker and Ordeshook 1973).

both American and Japanese, had the background competence to grow in that theory-inflected department.

> Lionel realized that . . . the Economics Department would always be small by the standards of the profession. . . . Most heads of department would therefore aim for a program that focused on a particularly narrow set of fields. . . . Lionel's view was different. . . . [H]is aim was to have our program reach out to many applied sub-fields of our discipline, but by hiring scholars who not only had a respect for basic theory but who knew how to apply that theory to their own applied sub-field. (Jones 2012, 11)

This departmental focus was less congenial, in the late 1950s, to older economists than to younger ones. Consequently McKenzie recruited at the junior level. He had remarkable success hiring such future stars as Stan Engerman, Sherwin Rosen, Rudiger Dornbusch, Michael Mussa, Buz Brock, Marcus Berliant, and Paul Romer. Most of these individuals eventually left Rochester for richer and larger departments, but the excitement of being in such an active research community was long-lasting.

> Lionel was not a "friendly buddy" to young recruits, even as they aged. He was a "loner" in many ways. My guess is that this was probably reflected in the names used to address him by different groups. I think all of us on the faculty called him "Lionel." However, the friends and associates who knew him before his Rochester days always seemed to call him "Mac," even Blanche called him that. . . . [M]ost of his former students [called him] "Professor McKenzie." . . . I often thought of him as a "security blanket." (Jones 2012, 12)[15]

---

15 A former student, Randi Novak, recalled that "[when he attends] economic presentations, he hears everything and remembers everything and understands everything, from details to the big picture. Dr. McKenzie waits until all attendees have spoken. One may think the presentation is over, but in his usual gentlemanly manner Dr. McKenzie asks one last question that references a small error that nobody picked up on that brings down the entire theory of the presenter" (personal communication).

At the end of the 1950s, McKenzie's work took a final turn after he read the 1958 book by Dorfman, Samuelson, and Solow on linear programming and its discussion of economic growth in the framework of linear models. John Hicks had come to Rochester in spring 1960 to give a talk on economic growth, the same talk he gave in both Berkeley and Japan.[16] This talk led directly to the so-called turnpike theory with seminal contributions from Roy Radner at Berkeley, Michio Morishima in Japan, and McKenzie, each of whom had heard Hicks's talk.

Debreu never wrote about growth theory. He believed that such dynamical economic models could never be analyzed since there were no natural dynamic adjustment mechanisms. Since there were no economic laws that could be described by differential equations, there could never be agreement on the structure of any dynamic model. Out of equilibrium nothing, or anything, could be said. With no theory of dynamics, no study of dynamic models could ever be sensible, let alone rigorous (Weintraub 2002, 145–46). Between Debreu and McKenzie, there was little common ground.

And yet McKenzie's career, and Arrow's and Debreu's, were linked in the minds of others sophisticated enough to see who was on the theory frontier, such as Samuelson:

> I know rather well most of the leading theorists in the world; and I have no hesitations that McKenzie belongs in their very top ranks. Indeed he is quite unusual in that he has a deep understanding of the *economics* of mathematical economics: only K. Arrow of Stanford, R. Solow of MIT, and possibly T. Koopmans of Yale would I put above him in this regard. At the same time he is right up at the frontier of Debreu-type mathematics. (Samuelson to A. Osther, Assistant Officer Oxford, 1961, June 28, PASP, 51, McKenzie)

---

16 This likely was the early version of his "Prices and the Turnpike: The Story of a Mare's Nest," which appeared in the *Review of Economic Studies* in 1961 and was itself a marker on his road to *Capital and Growth*, which appeared in 1965 (see McKenzie 2012).

## Arrow's Preeminence

Arrow's career subsequent to the *Econometrica* submission in summer 1953 ratified the faith that Hotelling and Wald had placed in his abilities. He was both prolific and eclectic in his tastes. Over the two decades before he was invited to Stockholm, Arrow would write close to one hundred articles, a large share of which were coauthored with, among others, Armian Alchian, Marc Nerlove, Frank Hahn, Mordecai Kurz, Robert Solow, Hirofumi Uzawa, Herbert Scarf, Roy Radner, and Samuel Karlin. He cooperated with many individuals particularly in connection with projects undertaken at the RAND Corporation. He wrote only one paper with most of his collaborators, but with Leonid Hurwicz (who, as we have seen, was openly critical about Debreu's work) he coauthored more than ten articles. Arrow was so embedded in a large and open community that he seemingly lived unconcerned with his status and role. His early omnivorous reading created numerous opportunities for connections with disparate intellects, and he sought opportunities to escape the narrower economics communities of like-minded people.

Arrow's career was thus not much shaped by the existence proof. In contrast to Debreu and McKenzie, he did not continue working on the existence of equilibrium for competitive economies. In the introduction of his article with Debreu he clearly stated that the existence proof was no more than a preliminary exercise "for descriptive and for normative economics" (Arrow and Debreu 1954, 265). The existence proof permitted the examination of other questions, but it had no value in and of itself. Post-1954 he lived out his "descriptive and normative concerns" in a mix of fields spread among behavioral economics, operations research, and applied microeconomics.

The first and most urgent question in getting closer to the "descriptive and normative purposes" of general equilibrium theory involved the classic issue of Walrasian economics, the stability of equilibrium: If the system were in a nonequilibrium state, would the dynamic laws of the system move the system back to an equilibrium state? If prices are the states of the competitive markets system, would the dynamic "laws of supply and demand" suffice to drive prices to the competitive equilibrium? The process, called by Walras the *tâtonnement*, is not a part of any existence argument, as Arrow noted (Arrow and Debreu

1954, 266). Stability is essential with regard to the central intuition of equilibrium analysis, namely that equilibrium can be achieved by market forces. Arrow's "fictitious player," whom Debreu attempted to exclude, alluded to this process.

Arrow joined forces with Alain Enthoven (1956), a graduate student at MIT and later RAND scholar, and with Marc Nerlove (teaching at Stanford 1960–65). He wrote the most important papers on this topic jointly with the mathematician Henry David Block and Leonid Hurwicz (formerly at Cowles), both at the University of Minnesota (Arrow and Hurwicz 1958; Arrow, Block, and Hurwicz 1959). Hurwicz twice, in 1955 and 1958, had a visiting position at Stanford. Reminiscing about his collaboration with Arrow, Hurwicz commented:

> [H]e is an ideal collaborator and willing to carry, if anything, more than his share of the work effort. As for giving credit to others (and I don't mean just to his collaborator, but to others who have worked on a problem), I do not know of another person who is more generous and who magnifies the accomplishments of his predecessors. . . . He is really punctilious about searching out anything in the literature that might have preceded his own contribution. (Hurwicz, in Feiwel 1987, 663)

It was Samuelson's 1938 paper, reprised in his *Foundations of Economic Analysis* (1947), that set out the stability problem in its modern form. The revived interest in stability during the mid- to late 1950s was associated with increased awareness and appreciation for various preexisting mathematical tools. This new mathematical economics literature employed the difference between local and global stability, introduced economists to the Lyapunov Theorem on stability, brought in Samuelson's arguments about the so-called weak axiom of revealed preferences and the strong assumption of gross-substitutability, and ultimately ended in a dead end of systematic counterexamples to the conjecture that there are necessary conditions ensuring that the competitive equilibrium will be stable (e.g., Scarf 1960; Gale 1963). There appeared to be no possible unitary and rigorous description of how market forces work.

While McKenzie clearly shared Arrow's interest in stability, Debreu remained uninterested. Werner Hildenbrand recalled that with respect to Hurwicz's and Arrow's work Debreu merely "shook his head. He knew at the outset that this leads to nowhere" (personal communication). For Debreu equilibrium was a matter of the consistency of a model and not a state of the world: "when you are out of equilibrium," he later explained, "you cannot assume that every commodity has a unique price because that is already an equilibrium determination" (Debreu, in Weintraub 2002, 146). Disequilibrium, for Debreu, was a logical contradiction. Arrow recalled in a letter to Weintraub dated September 19, 2010:

> Leo Hurwicz and I wrote a paper on stability of equilibrium. The referee took almost a full year though he did not have any very drastic comments. I later learned that Gérard had been the referee. His refereeing did stimulate, however, some very important papers of his own, on structural stability of Walrasian equilibrium,[17] so perhaps the invisible hand was working (though Leo and I were not happy).

Even if stability theory proved to be a dead end, it was hardly the end to Arrow's verve and economic research. One of the fields he tackled, somewhat continuous with the idea of how economies change and adapt, was his work on learning, which later became relevant for what came to be called endogenous growth theory (1962). He published on risk-taking, moved beyond the notion of equilibrium in rationing theory in disequilibrium situations, and shared with Hurwicz a strong interest in organizational and informational issues. At the same time, he was exploring at a more basic level ideas of nonlinear programming that had been developed at the First Stanford Symposium "Mathematical Methods in the Social Sciences." He organized that meeting in 1959 with a colleague from the psychology department, Patrick C. Suppes, and a Japanese student whom Arrow had recruited to Stanford, Hirofumi Uzawa (1960). What appear to be disparate topics are in fact mutually coherent if one notices Arrow's

17 Structural stability refers to the idea that as the model or system changes, the topological character of the set of equilibrium solutions remains unchanged. This differs from "stability," which inquires about the ability of the model or system to return to an equilibrium state if disturbed. The latter requires a theory of dynamics, the former does not.

interest in behavioral issues in the context of operations research. This research program shaped by the RAND Corporation embedded Arrow's work in a broad array of social science disciplines. That the program had its beginnings in the vision of cybernetics developed in the early cold war period was not of course to limit its applicability to a wider range of problems and interests. Arrow was an important part of Stanford's development as an academic stronghold of this vision of science.[18] He was less a figurehead at Stanford than was Debreu in Berkeley in his group or McKenzie in the economics department at Rochester but was rather a link among the departments of economics, operations research, and statistics.[19] In the economics department Arrow was welcomed by the chair, Bernard Haley, and faculty members Lorie Tarshis, Tibor Scitovsky, Moses Abramovitz, and Mel Reder. Scitovsky would be most sensitive to Arrow's work in social choice. Shortly after Arrow's arrival, Stanford hired Hollis B. Chenery and Hendrik Houthakker, both of whom later would be hired by Harvard. With Mordecai Kurz in 1959 Arrow created the Institute for Mathematical Studies in Social Sciences, and that group came to be known for its summer meetings (today it is called the Stanford Institute for Theoretical Economics). Arrow recalled: "I became rather influential rather quickly, partly because I was very conscientious and read everybody's manuscripts and discussed their work. We were very careful in appointments and promotions. . . . There was a lot of vitality and life about the place. We had very few graduate students" (Arrow, in Feiwel 1987, 650).

In contrast to McKenzie at Rochester and Debreu at Berkeley, students were not the most important channel through which Arrow's influence grew.[20] Next to reading his colleagues' work in progress and

---

18 For this reason, Arrow was identified (next to von Neumann) as the main protagonist in Mirowski's (2001) and Amadae's (2003) politicized accounts of the transformation of economics during the postwar period, accounts that argue that actors like Arrow were cold-warriors at heart, or dupes of the cold war establishment. It should be clear that our history has little connection to, and even less interest in, such good guy–bad guy political arguments.

19 Stanford in the early 1950s was not what it is today. The emergent wealth of the entertainment and aerospace industries in Southern California were not yet matched by the later riches of Silicon Valley. The physics and psychology departments certainly had excellent reputations.

20 Nevertheless, among the students he supervised there were important economists like Karl Shell, Steve Goldman, John Geanakoplos, Jerry Kelly, Eric Maskin, Roger Myerson, Daniel

giving detailed comments, his influence grew primarily through his building relationships beyond the economics department in Stanford. He connected with the psychology department via Patrick Suppes, a link that was vital for the social science network in Stanford. His collaboration with Samuel Karlin (at Stanford after 1956), an eclectic mathematician active equally in molecular biology as well as in game theory, proved to be important in the extended interdisciplinary networks. With these contacts, Arrow always had one foot in the Center for Advanced Study of Behavioral Science in Palo Alto, which had been created in 1954 with a grant from the Ford Foundation. The organizational model for the center was the one already established at the Institute for Advanced Study in Princeton. Each year about thirty visiting scholars from several disciplines (anthropology, economics, political science, psychology and sociology, linguistics, and biology) were invited to collaborate in a synergetic fashion. Arrow was at home in this interdisciplinary world (see Lowen 1997).

Herbert Scarf, assistant to Patrick Suppes, had received a mathematics Ph.D. from Princeton in 1954 and worked at the RAND Corporation from 1954 to 1957 before joining the Stanford statistics department faculty in 1957. He, too, was to be a fellow of the center in 1962–63 before moving to Yale and Cowles in 1963.

> The atmosphere at Stanford was genuinely exciting. There was a sense of the great potential utility of mathematical reasoning in a variety of novel areas: in mathematical biology, statistical decision theory, game theory, and in mathematical economics. Arrow had completed his work on social choice, and had collaborated with Gérard Debreu on demonstrating the existence and welfare properties of a competitive equilibrium, with a generality and elegance made available by the theory of convex sets. Arrow was now working in the stability of equilibrium, which he was investigating jointly with Hirofumi Uzawa and Leo Hurwicz. It was this subject which I turned to as my first research topic in economic theory. (Scarf 1991, 127)

---

Osborn, Andrea Prat, and in Operations Research Harvey Wagner. Maskin and Myerson of course were future Nobel Laureates.

Together with Samuel Karlin, Gerald Lieberman, Herbert Scarf, and Harvey Wagner, in the mid-1960s Arrow launched yet another soon to be influential group at Stanford, the Operations Research group. Beginning as a Ph.D. program after George Dantzig arrived in 1966, the Department of Operations Research was formally established in 1967 with Lieberman at its head. Other active members were Daniel Teichrow, Charles Bonini, Frederick Hillier, Roy Murphy, Arthur Veinott, Peter Winters, Ronald Howard, Robert Wilson, Alan Manne, and Rudolf Kalman. These intellectual associations fostered and supported Arrow's interest in a variety of applied subjects such as the management of water resources (Davis 1990, 2) and health economics (Arrow 1963), the latter bringing him back to the world of actuaries. To be sure, many of these kinds of applied topics would be brought to him by others, but he was the one to pick them up. When his interviewer George Feiwel asked him what had led to his work in applied economics, Arrow answered:

> Primarily two factors: social conscience and an attempt to be a complete economist. Somehow I always had this idea that one ought to be socially useful . . . when we talk about studying people, we also talk about advising them. This dualism is very deep in the subject. (Arrow, in Feiwel 1987, 220)

This dualism of "descriptive and normative purposes" recapitulates his interpretation of equilibrium analysis. He was unsympathetic to the manner in which such analysis was increasingly used in economic research; the hermetic spirit of such analyses stood in stark contrast to his open interdisciplinary-cybernetics spirit. As Hurwicz commented on Arrow's sense of what it meant to do applied work:

> [H]e has become convinced that in some cases the problem is not so much lack of theory (although that may be also a problem), but rather that the theory already in existence is ignored or misused in the applied world. In particular, people often use the Arrow-Debreu theory of perfectly competitive markets precisely where it is not applicable. It is very interesting that Arrow's applied work has been just in those areas where the competitive theory in not applicable. . . . I think this direction

> of activity may arise from a conviction that people should be prevented from using the Arrow-Debreu model. (Hurwicz, in Feiwel 1987, 664)

Cambridge's Frank Hahn shared this view, and their magisterial *General Competitive Analysis* (1972) established Arrow's position on interpreting general equilibrium analysis.

In 1967 Arrow was invited to accept a chair at Harvard. As it was becoming known that Arrow might consider leaving Stanford, he was also offered a distinguished position at MIT. This was a clear sign of the changes that had taken place since the early 1950s when a hire like Arrow would have been impossible at Harvard. What was at the margin now had become mainstream, and Arrow was one of the establishment's central figures. After Arrow accepted Harvard's offer, Samuelson (Arrow's sister's brother-in-law)[21] welcomed him to Cambridge. Arrow replied:

> Dear Paul: I appreciated your kind words of welcome to the Cambridge community . . . [under] any auspices, and I look forward eagerly to seeing much more of you. I think you know that were it not for the presence and leadership of you and Bob at Cambridge, I would not be leaving Stanford . . . I think I have already indicated to you the motives for my decisions. Primarily I feel more at home in a university in which other social sciences and the humanities are well represented. . . . I also feel that there is something very useful I can do at Harvard that simply doesn't need doing at MIT. (PASP, 12, Arrow)

But Arrow would never really feel at home at Harvard, though he would stay for twelve years before returning to Stanford in 1980: "[O]ne felt even at dinner parties that one was in the midst of discussion of things that would affect the nation," he commented later on the rather ridiculous pretensions present at Harvard (Arrow, in Feiwel 1987, 652).

---

21 Arrow's younger sister Anita, also an economist, had married Samuelson's younger brother, the economist Robert Summers.

> I had some trouble for a while getting another theorist. By hiring me they thought that had done all they should do for theory, for traditionally the department had been anti-theoretical. There was no question I was employed because they felt that the world felt that they were backward because they lacked a theorist. (ibid.)

One suspects that his use of "for a while" refers to his first four years at Harvard because in 1972 his academic fame grew larger as he received the highest public award that could be given to an economist: the Nobel Prize in memory of Alfred Nobel.

# CHAPTER 8

## THE PROOFS BECOME HISTORY

By the early 1960s our protagonists had found stable intellectual homes. Debreu contributed to the growing strength of the Berkeley economics department, McKenzie built a theory-inflected economics department in Rochester, and Arrow, after having built up a network including economics, statistics, and operations research at Stanford, moved to a Harvard that once had eschewed Jewish scholars. Although independent one from another, credit for the existence proof given to one of them, and not another, would disturb their professional equilibria.

It is a truism in intellectual history that creative works, in gaining significance through their use by others, develop a life of their own. The context in which a piece of work is produced is largely disconnected from the paths through which it becomes influential. Tracing the paths along which the existence proofs of the early 1950s became influential would involve an immensely complex narrative surveying nearly all research fields in economic theory. What we can say is that by being applied, interpreted, shaped, and reshaped, the proofs would soon come to symbolize a new intellectual culture in American economics. The proofs helped reconstruct the body of economic knowledge. But, more important, the new image of this mathematical economic knowledge stabilized the disciplinary boundaries of economics. In this new regime of truth-making, economic knowledge was to be impersonal and austere.[1]

---

1 This distinction between the body of knowledge and the image of knowledge was given prominence by the historian of science Leo Corry (e.g., 1989, 411), who credited Yehuda Elkana with the original idea. This distinction was the organizing trope in Weintraub 2002.

The gap between the context of origin of a piece of work and the context of its later representation has always inspired historians: the separation permits our conceiving the very historicity of human artifacts. Describing this difference usually results in a hierarchical order valorizing the origin but not the representation of a work, thus producing narratives in which intrepid pioneers introduce new modes of thought and gain adherents as they overcome obstacles. Whatever the social function of such narratives, they are not our concern. However one may describe and evaluate this postwar transformation of economic theory, what we have seen in the preceding chapters is that none of our protagonists intended to produce this change—instead they lived through it. Arrow, Debreu, and McKenzie were not driven by a clear vision of the foundational role of their proofs for the discipline of economics at large.

In this chapter we describe questions of credit generically, as emerging from the reception of our protagonists' work. We will catch only a glimpse of the multiple and complex narratives of the influence of the equilibrium proofs. The decisive issue for us is that the more influential their work became, the more our protagonists had to accommodate to the fact that they, personally, were the ones who would be credited and held responsible for the transformation of the intellectual culture in economics. It was only during the years that followed, in which the significance of the postwar period became apparent to economists beyond the membership of the Econometric Society, that economists asked: "Who should earn the credit for effecting these changes?" Issues of historical agency became pertinent for our protagonists through questions of scientific credit. Our own story, in the pages that follow, thus reverses the order of origin and representation. That is, we do not try to understand the history of these ideas by contrasting the contexts of origin and influence. Instead we seek to understand the history of these ideas recognizing that the award of scientific credit creates questions of agency in the first place. Our work is one step forward in the project of historicizing general equilibrium theory since we deconstruct, as it were, the internal accounts of general equilibrium theory in the history of economics that are shaped by its celebration. Our narrative will thus have a reflective dimension: as with the patency of credit, our protagonists developed personal stakes in the historical reconstruction of the postwar transformation of economics particularly after the creation of the first "Nobel" Prize for economics in 1969. Put another way, we ask what it meant to Arrow, Debreu, and McKenzie to be acknowledged, neglected, and misunderstood as the agents of the transformation of economic theory.

## The Influence of the Arrow-Debreu Paper

Credit both precedes and follows influence. As the three men's career paths differed, neither the arcane purity of Debreu, nor the applied eclecticism of Arrow, nor the theoretical predilections of McKenzie represented channels through which the existence proofs became influential in the profession at large. Without reviewing the complex history of the reception of their paper, we can loosely identify three paths along which "Arrow-Debreu" traveled: first, through its applications in fields different from those in which our protagonists were active; second, through the re-politization of equilibrium analysis by its critics; and third, through its uses for the teaching of economics.

Within the growing though still small community of economists who regularly read *Econometrica*, the several existence proofs quickly became a model of rigorous research. In the larger community of economists, however, it took a decade or so before the conditions under which a general competitive equilibrium could be established became common knowledge. The influence of "Arrow-Debreu 1954" in economic research is well-documented by citations of the article.[2] During the first ten years, there were thirty-six references, most of them in *Econometrica* from colleagues of either Arrow or Debreu. Until Arrow's Nobel Prize in 1972, roughly the same number of papers referred to the article, though an increasing number had a computational focus. Historians were surprisingly quick to proclaim the historical importance of the general equilibrium existence proofs. During the 1960s there already was a reference to the article by an historian of economics (Jaffé 1967). Jaffé de-homogenized Menger, Jevons, and Walras but contributed to the homogenization of Arrow

2 Based on data from the Social Science Citation Index and Google Scholar. For similar results, see Oehler 1990.

and Debreu.[3] From 1972 until Debreu's Nobel Prize in 1983, references increased to more than eighty, many of them from macroeconomics and neighboring fields such as finance, monetary theory, international trade, and even regional studies. Citations continue to appear to the present day though, to be sure, references in *Econometrica* have decreased. Even if we consider the substantial growth of the profession during these several decades, the article became a benchmark of research *no sooner* than the mid-1970s in subdisciplinary areas that neither Arrow nor Debreu considered to be their primary audiences.

One important channel through which the proof became influential was in what was then called "the microfoundations of macroeconomics" debate in the period of the late 1950s through the 1970s. Arrow-Debreu became the central trope of this debate and further displaced the equilibrium papers from their context of origin. Briefly, even as Samuelson was announcing the birth of the "neoclassical synthesis," many theorists tried to clarify the conditions under which microeconomics and macroeconomics could be made consistent. Intuitively, there should be just one economic theory, and particular specifications should produce both neoclassical equilibrium and Keynesian models. Patinkin's discussion in his famous chapter 13 of *Money, Interest and Prices* (1956), based on his book's "Note B: Walras' Theory of the Tatonnêment," identified Keynesian involuntary unemployment phenomena as disequilibrium "errors" in the competitive equilibrium model.[4] Subsequent work by Robert Clower, Frank Hahn, Axel Leijonhufvud, Bent Hansen, Franklin Fisher, and others, surveyed by Weintraub (1977), employed "Arrow-Debreu" as metonymic for microeconomic theory and "Keynes" as metonymic for macroeconomics. It was only then that Arrow-Debreu was tagged with a political meaning that stood in contrast to the historical baggage of the early postwar period that it still carried, for example,

3 Considering the increasing interest in reconstructing the history of general equilibrium theory among historians during the 1970s and 1980s (Hutchison 1977, Blaug 1980, and Weintraub 1985 to name but three different approaches), it is interesting to speculate about how much they contributed to the identification of Arrow-Debreu with rigorous economics.

4 In his first edition of *Money, Interest and Prices* (1956), Patinkin referred to Wald and Schlesinger and von Neumann.

among the pro-planning "French engineers." This rhetorically corrupt equation of Arrow-Debreu versus Keynes and market efficiency versus inefficiency led to the conclusion that neither post-Keynesian nor neo-Austrian economists could learn anything from the mainstream texts of Varian and Mas-Colell. If one is socialized to the economics community by means of acceptable textbooks, as in a Kuhnian normal science ([1962] 1996, 20), economists whose concerns cannot be taught by the field's textbooks must necessarily be heterodox with respect to the mainstream. Such heterodox economists produced the litany of popular complaints regarding the overuse of mathematics, for which Arrow-Debreu became *the* bogeyman. For an entire community of post-Keynesian and otherwise heterodox economists, lamenting the limits of Arrow-Debreu became constitutive for their very identity as economists (see Backhouse and Boianovsky 2013).

Textbooks provide another indicator of the influence of the article. The folk knowledge that textbooks lag behind research and report results only after they have been well established does not apply to Arrow-Debreu 1954. By 1958 the proof had been mentioned by James Henderson and Richard Quandt in their widely used microeconomics textbook[5] and was fully described in texts even before Arrow's Nobel Prize in 1972 (see Weintraub 2002, 188). If graduate textbooks shaped the socialization of new professional entrants, those books' presentation of the Arrow-Debreu theorem became the canonical benchmark of rigor. The books served to raise the entrance barriers into the economics profession. This channel of influence remained stable into the 1990s: Varian's text of 1978 and of course Mas-Colell, Whinston, and Green's of 1995 taught Arrow-Debreu to the professional entrants. The Arrow-Debreu paper thus did not simply enter the profession's consciousness as a research result. Having its origins in the institutions of mathematics, it developed a life of its own in constructing borders between disciplines.

Without probing any deeper, the preceding sketch of how the equilibrium papers became influential makes it clear that, in con-

5 Henderson was at the University of Minnesota, and Quandt was at Princeton. Their textbook appeared in the important McGraw-Hill Economics Handbook Series.

trast with the simple narratives our protagonists construed in their own communities, the issues of who is to be credited or blamed were quite complex. It is now sixty years after the events, and thirty years after the first historical exploration of those events (Weintraub 1983). Time provides perspective. The previous chapters, based as they are on the present availability of archival material, allow us to present a generic account of issues of credit as they intruded on the career paths of our protagonists. This genealogy allows us to explore the difficulty of representing a person or persons to be credited for the postwar change in economic theory. By the 1980s the existence of equilibrium papers had become more than contributions to a community's body of knowledge. They were becoming exemplars of a change in the image of what economic knowledge might be. In this role the papers, and their authors, entered history. They were to be appraised not solely in terms of their links to other papers and ideas, as in normal scientific practice, but as evidence that economics itself had changed. The creation of the Bank of Sweden Prize in Economic Sciences in Memory of Alfred Nobel in 1969 confirmed precisely this post–World War II scientific nature of economics. It also introduced something to the community of economists that had not been there before: symbolic public credit.

## Arrow's Nobel Prize of 1972

The creation of the Nobel Prize in 1969 was meant to recognize the established place of economic science in Western society. To be sure, the creation of the post of Chairman of the President's Council of Economic Advisers in 1947 "elevated" an economist to public office, and the Clark Medal of the AEA, first given in 1947, reified professional achievement. But the Nobel Prize was different. Economists would stand shoulder to shoulder with chemists, physicists, doctors, writers, and peacemakers in front of the Swedish king.

In September 1968 the Swedish Royal Academy of Science wrote to Paul Samuelson asking him to suggest worthy Nobel candidates. Samuelson named fourteen economists over the age of sixty (of whom six would eventually receive the prize) and twelve younger economists (of whom seven would later receive the prize). He named

Michał Kalecki and Leonid Kantorovich as potential candidates from Eastern Europe (PASP, 36, Hansen).[6] But before the prizes would be given to those on Samuelson's list, the second award (after Frisch's and Tinbergen's shared prize in 1969) would go to Samuelson himself. In the years to come individuals associated with the Cowles Commission would receive more than half of the first dozen Nobel Prizes. The fourth prize in 1972 went to Kenneth Arrow, jointly with Sir John Hicks.

This would be the first major confusion regarding public credit. The specific press release claimed that the award was "for their pioneering contributions to general economic equilibrium theory and welfare theory." The ensuing discussion of Arrow's work mentioned that "the pioneering work, a paper from 1954, was written together with Gerhard [*sic*] Debreu." It went on to state that the model presented in their paper became the starting point for further research in this field. The discussion then moved off to discuss Arrow's work in welfare economics associated specifically with the fundamental theorems of welfare economics and the Pareto efficiency of the competitive equilibrium, but there was no mention of Arrow's contribution to social choice theory. Although the criteria that inform the decision of the committee might be contingent and obscure, it is curious in any event that Debreu was described in terms similar to that of Arrow but that there was no prize for him (even if his contribution to welfare theory was limited). And there was no mention of the role of Lionel McKenzie.

What an irony indeed coupling Arrow with Hicks! As we have seen (chapter 5), one of the historical conditions for the existence proof to have become resonant was a preference for a kind of mathematics that disqualified John Hicks as a rigorous economist. One might argue that Arrow and Hicks did share some sense of eclecticism in

---

6 Samuelson listed Frank H. Knight, Alvin H. Hansen, Jacob Viner, Jan Tinbergen, Ragnar Frisch, Bertil Ohlin, Gunnar Myrdal, Roy Harrod, John Hicks, Joan Robinson, Wassily Leontief, Gottfried Haberler, Nicholas Kaldor, and Abba Lerner of the older generation and Lloyd Metzler, James Tobin, Robert Solow, Kenneth Arrow, Franco Modigliani, Maurice Allais, Tjalling Koopmans, Milton Friedman, Erik Lundberg, Ingvar Svennilson, Hermann Wold, and Henri Theil of the younger generation. Debreu and McKenzie are not on the list. Thanks to Yann Giraud for directing us to this folder.

their work, but they did not share anything for which they were receiving credit. Hicks did not contribute to the transition in economics that shaped the postwar profession. Except for *Value and Capital*'s modest discussion of temporary equilibrium, his work would play virtually no role in post-1954 developments in general equilibrium analysis (even if postwar theorists all had read Hicks in their youth). Morgenstern's hostile 1941 review of *Value and Capital* in the *Journal of Political Economy* was a harbinger. If his book had helped form the taste of an entire generation of economists, including Arrow, in the immediate prewar period that generation was to find a new vocabulary and grammar by the early 1950s. Moreover Hicks himself, even as early as 1972, was in the process of repudiating, or at least backpedaling away from, his 1939 book and opposing the intellectual frameworks of both IS-LM analysis and general equilibrium theory.

How must have McKenzie felt about the prize? It must have been confusing, considering his mixed feelings about Hicks as his Oxford supervisor. Moreover, Debreu was mentioned in the award to Arrow while he was not. And Debreu? How did he feel about Arrow's prize? Hildenbrand recalled:

> [Arrow's Nobel] was rather disadvantageous for Debreu, since with Arrow equilibrium theory was covered. One could not expect that it would be chosen for a prize once more. Debreu was rather tense in these years. He no longer believed that he would get the prize. That made him nervous. Of course he thought he would get it jointly with Arrow. But that they packed Arrow and Hicks together—that was a joke. (Werner Hildenbrand, personal communication)

The symmetrical lack of Nobel Prizes to Debreu and McKenzie contributed to both a sense of competition and identity between the Berkeley and Rochester economics communities. The strong social bonds at Rochester described in the previous chapter reinforced the sense of a compact intellectual community determined not to be seen as a minor league Cowles or, worse, a second-rate Berkeley. The competition between Debreu and McKenzie, invisible in the public record, was a constitutive bond among McKenzie's students once the Arrow-Debreu proof entered "the textbooks" rather than McKenzie's own proof, which was canonical at Rochester. Away from the public eye, this competition with the group in Berkeley can be inferred, for example, from a McKenzie letter to Samuelson dated June 17, 1975, discussing Hal Varian's comments on one of McKenzie's articles (PASP, 51, McKenzie):[7] "I guess Varian's point is related to the new uniqueness theory of Dierker, etc. (all those Berkeley characters). It is an obvious lift from John Milnor's book *Topology from a Differentiable Viewpoint* (which I had around for years without reading, dammit)."

This fraught relationship between McKenzie and Debreu could be awkward for others. Marcus Berliant was a 1982 Berkeley Ph.D. whose general equilibrium theory dissertation was supervised by Debreu. He took up a position as an assistant professor of economics at Rochester in July 1982 before he had formally defended his thesis. Berliant's own Ph.D. students were told the story that has been part of the folklore for years at Rochester: when Berliant went back to Berkeley to defend his thesis, in which he referred to the Arrow-Debreu-McKenzie model, Debreu refused to allow him to take his orals until he removed the references to McKenzie (Thomas Nechyba, personal communication).

Arrow, instead, remained generous. Responding to a congratulation letter from Debreu, he wrote on November 2, 1972: "I think of the prize as a recognition of the whole development of rigorous quantitative thinking in economics to which you and a few others have been the major contributors. I'm glad that the citation specifically recognized our joint paper of 1954" (GDP, Carton 5, 102).

## Agents Become Historians

The publicity surrounding Arrow's Nobel Prize increased competition between the Berkeley and Rochester communities, and so Debreu and McKenzie had now greater personal stakes in reconstructing the history of their 1950s work. They had several occasions to present their own accounts and thus "make" history. At the fiftieth anniversary celebration of the Cowles Commission in 1983, Debreu

---

7 Varian was a 1973 Berkeley Ph.D. who had taken up a first faculty appointment at MIT.

presented an historical account of the changes in the 1950s. Without mentioning names, he reconstructed the events that gave space for "crediting" his own contribution in representing mathematical rigor.

> With the fifties began a new phase in the development of the theory of general economic equilibrium, which was due for a thorough reexamination. The first problem to be investigated was the characterization of the Pareto optima of an economy by means of a price system. The many contributions made to this subject since the end of the nineteenth century had relied on differential calculus. . . . As a consequence, those contributions did not satisfy the new, more exacting standards of rigor prevailing in economic theory. It now turned out that the traditional, troublesome assumptions of differentiability were superfluous, and that convex analysis, in particular the supporting hyperplane theorem, provided a rigorous answer that was more general, more natural, and simpler. (Debreu 1983a, 30–31)

Note that by emphasizing the welfare context of general equilibrium theory, even if unwittingly, Debreu "wrote out" McKenzie.

One of the occasions at which McKenzie reconstructed his version of the same events was his presidential address to the Econometric Society in Ottawa and Vienna in 1977, a revision of which was published several years later in 1981. In it he set out to

> discuss the present status of a classical theory on existence of competitive equilibrium that was proved in various guises in the 1950s by Arrow and Debreu, Debreu, Gale, Kuhn, McKenzie, and Nikaido. The earliest papers were those of Arrow and Debreu, and McKenzie, both of which were presented to the Econometric Society at its Chicago meeting in December, 1952. They were written independently. The paper of Nikaido was also written independently of the other papers but delayed in publication. (McKenzie 1981, 819)

In conversations, McKenzie would repeat this account over and over again, and it would become a critical narrative for the Rochester community, for example at the Wednesday afternoon tea in the

Rochester economics departmental lounge, which McKenzie continued to attend regularly even after his retirement in 1989. Roman Pancs recalled:

> Several themes were recurrent in Lionel's Wednesday afternoon tea-room conversations: that Lionel was delighted that Arrow acknowledged his priority in proving the existence of the general equilibrium; that Debreu did not cite Lionel's paper and did not tell Arrow about it; that at the meeting of the Econometric Society, Debreu said that Lionel's paper was a special case of Arrow and Debreu's paper, to which Lionel retorted that, perhaps the opposite was true; in fact, neither paper was a special case of another. (Roman Pancs to Weintraub, October 4, 2011)

The sense of being underrated pervaded McKenzie's professional identity. In a memorial to McKenzie, David R. Henderson, an assistant professor at Rochester between 1975 and 1979, reported another episode with similar emotional content:

> He was also classy even when treated in a non-classy way. The last seminar I went to at the University of Rochester was in May 1979, just before I left for a job at the Cato Institute in San Francisco. The speaker was Paul Samuelson who, as I recall, was there to receive an honorary doctorate. Lionel introduced Samuelson and reminisced graciously about how Samuelson had recommended him when the University of Rochester had asked who would do a good job of starting the Ph.D. program. Then Samuelson started to speak. I'll never forget what he said. He said that back then he thought Lionel was being undervalued in the market. "I'm not saying he was the best, mind you, but he was definitely being undervalued and I thought he could do a good job here." I think if that happened today, I would probably gasp out loud. Instead, I just looked at Lionel who looked hurt but quickly recovered and listened, apparently serenely. As I say, Lionel McKenzie was a classy man. (Henderson 2010)

McKenzie's autobiographical memory of that early 1950s period came to define both his professional and personal identity. The narrative of his grudgingly recognized priority, his struggle to find a place in economics, and his concern to receive proper credit shaped his persona. The professional frustration went hand in hand with his personal losses. Fred, his oldest son, had gone to fight in Vietnam, returned mentally broken, and later committed suicide. His second child, daughter Gwendolyn, developed breast cancer as young woman and, following what Lionel believed were medical errors, died prematurely. He wrote to Dipankar Dasgupta (2013), "Believe me, it's very hard to bear." His wife of fifty-six years, Blanche, died eleven years before him in 1999. He was not on good terms with his youngest son, David. A former student of McKenzie's recalled that he and another student were arguing about the value of a human life in the lounge one day:

> We had no idea that the venerable professor was listening to the discussion, so engrossed he seemed to be in the newspaper. It was a total surprise to us therefore, when, on his way out of the lounge after he had finished his coffee, Professor McKenzie suddenly turned around to face the group engaged in the discussion. And then, in a stentorian voice that rang through the department, he said: "Guys, let me assure you, human life's not worth a s***!!" Saying so, he guffawed and vanished into his office without waiting for a response. (Dasgupta 2010)

When McKenzie approached death at age ninety-one, he was still mentally revisiting the events of the early 1950s. In November 2008, he gave a last talk about the history of general equilibrium theory in Roman Pancs's graduate microeconomic theory class, still comparing himself to Arrow and Debreu: "Lionel mentioned that in some sense his model was more general than the Arrow-Debreu model, as it did not view firms as exogenously given" (Pancs to Weintraub, October 4, 2011). And on September 23, 2010, only a few months before he passed away, McKenzie mentioned to a student of Pancs the paper on which our reconstruction of McKenzie's contribution in chapter 6 is based (Weintraub 2011): "E. Roy Weintraub of Duke University has written an article defending my priority in proving existence of equilibrium but I don't know whether he will be able to get it published" (Pancs to Weintraub, October 4, 2011).

## Historians Become Agents

This was not the only time that the work of historians of economics was deployed in the historical reconstructions performed by our protagonists. In particular Weintraub's survey article in 1983, which treated Arrow, Debreu, and McKenzie symmetrically, both shaped and was shaped by the history of emerging credit for the early existence proofs. The symmetry attracted attention in the context of the rivalry between Berkeley and Rochester. It caused a sufficient stir that several years later when, in a 1989 letter to McKenzie, Samuelson suggested that he nominate Weintraub for the J. Murray Luck Award, an annual prize given by the National Academy of Sciences for Excellence in Scientific Reviewing (PASP, 51, McKenzie). McKenzie replied on May 22, 1989:

> I am happy to seek supporting letters for Weintraub, but I think it is more likely to fly if you are the nominator! I could write Leo Hurwicz and John Chipman, who received the first award. I could also write to Arrow. I doubt that Debreu would support the idea, but I could be wrong. I am not averse to approaching him, however. Maybe you could speak to Bob Dorfman. I don't know whether Bob Solow would think he was an appropriate person to write in light of the field. Maybe you could approach Larry Klein too. What do you think of this plan? (ibid.)

As we characterized Cowles in the early 1950s, it was a very small community. And it was the same coterie that would give credit to the historians they deemed to have gotten credit right. But on May 30, 1989, Samuelson replied to McKenzie: "On the Roy Weintraub nomination for the Luck Award, I am inclined to take the lazy way. It was a good idea but not the most pressing matter. Besides, as sluggards always say, someone [else] can take up the matter" (ibid.). Since "the matter" was dropped, it gives us reason to engage in some self-reflective reconsiderations of the contribution of the younger Weintraub to the emergence of the issue of credit for the existence proofs.

At this point, as in places earlier, the older of the two authors thus must step out of his authorial character. When the historian engages with the historical actors, he, too, becomes an historical actor. The historian and the protagonists all construct and reconstruct and criticize each other's accounts. The diction becomes awkward. Using "I" is inappropriate since Till Düppe and E. Roy Weintraub are the authors of this book. The only possible locution creates the historical personage "Weintraub" to which the authors, the "we," may refer and cite. Employing this third-person singular "Weintraub" to create distance from the authors offends the ear and appears to us to be both haughty and bombastic. Such issues are rife in constructing histories of contemporary science (Söderqvist 1997, 2003). Alas.

Giving credit is a form of historical judgment. It is thus only natural that historians of science have both an intimate and difficult relationship with those to whom credit is or is not given. Historians contribute to making these judgments by either justifying or criticizing them. The relationship between historian and scientist is even more difficult and intricate if the historian deals with questions arising from internal histories of science constructed not so differently from survey articles written by practitioners themselves. This tension was present in Weintraub's first contribution (1983) to the history of the early existence proofs. That paper was set in motion by his earlier survey of the emerging literature on the microfoundations of macroeconomics where, writing then as an economist doing a survey article not as an historian writing a history, he used "Arrow-Debreu-McKenzie model" to provide a reference phrase for the competitive equilibrium model, noting that "many authors use the term Arrow-Debreu model. Since, however, the *proof* of existence on which current work is based came out of McKenzie (1954), it seems appropriate to give McKenzie equal billing" (Weintraub 1979, 27n5).

When Weintraub wrote the 1983 article, Debreu and McKenzie were aware of his use of the phrase "ADM-model." As a neophyte historian, however, he was blithely unaware that the three protagonists were attempting to construct the historical record to accord with their own memories and interests. As long as the history of economics is written on the basis of the actors' account of their history, or as survey articles extended back in time, such confusions are unavoidable. In preparing his article in 1980–81 Weintraub had written letters to many individuals who had associations with the Cowles group: Kenneth Arrow, Gérard Debreu, Lionel McKenzie, Tjalling Koopmans, Gerhard Tintner, Carl Christ, John Chipman, Lawrence Klein, Herbert Simon, and Nicholas Georgescu-Roegen. He also wrote to Paul Samuelson and was in contact with Allen Wallis and Paul Wolfowitz, son of Jacob Wolfowitz, who was Abraham Wald's literary executor. The emergent story credited Arrow, Debreu, and McKenzie without raising issues of the appropriate share of recognition for each of them. The issues of priority, and the complex intertwining of problems of simultaneous discovery, lurked beneath the paper's surface, unappreciated by its author.

In one of the early responses to Weintraub's set of questions, Kenneth Arrow (November 19, 1981) mentioned McKenzie's priority in publication: "Meanwhile, McKenzie, working independently, had published his paper first, though it was somewhat less general." In his first response to the questions posed, Lionel McKenzie traced his own intellectual development and wrote, "I believe I was the first to use the Kakutani theorem this way, although I believe Nikaido's use of it in his paper in *Metroeconomica* (1956) was independent of mine. His paper was delayed in publication." He continued emphasizing the simultaneous presentation of their articles at the Chicago meeting where Debreu had spoken up in McKenzie's session: "My paper and the paper of Arrow-Debreu which were developed completely independently were presented to the December, 1952, meetings of the Econometric Society [in Chicago]."

A draft of Weintraub's paper (dated March 2, 1982) went to Arrow, Debreu, McKenzie, Koopmans, and Chipman. The first response (that we quoted earlier), concerning the public discussion at the December 1952 Econometric Society meeting, came from Debreu (November 19, 1981):

> I have no recollection of the episode recounted . . . and I cannot testify one way or the other on this matter. I bring this question up because you might have interpreted absence of comment on my part as an endorsement of the statements that you quote. T. C. Koopmans may possibly remember what happened at that session. . . . Another point must also be noted that ac-

> cording to the [Weintraub] account Lionel had not attended the seminar where I spoke and had no knowledge of the Arrow-Debreu paper. It's stated the next day that [he said] his paper implied our result.

McKenzie's response to the draft (April 16, 1982), raised a new point:

> [with respect to Arrow and Debreu's references to Wald] I assume that the remarks on page 289 of [the] Arrow-Debreu existence paper about the weak axiom were meant to imply that the weak axiom was used to get uniqueness but was not depended on for existence. [This refers to Wald's use of the weak axiom of revealed preference.] Reading their remarks in retrospect one would have thought that they understood the special character of the theorem in view of the assumptions, but they may not have read the [Wald] proof closely!

In other words McKenzie was shocked that Arrow and presumably Debreu were not aware that Wald's use of an assumption tantamount to the weak axiom of revealed preference was essential in Wald's proving the existence of equilibrium. It was the case, as McKenzie argued, that this disguised weak axiom created a world in which there was effectively but one consumer, which made the problem of existence quite simple. McKenzie was thus claiming that Arrow's discussion of Wald's paper meant that Arrow was unaware of that issue.[8]

Soon after the publication of the article, it attracted the interest of those creating credit, specifically as it preceded by only months the announcement of the Nobel Prize to Debreu. When Robert Dorfman was asked to write an essay for the *New York Times* following Debreu's Nobel Prize (Dorfman 1983), he wrote to Weintraub apologizing for an omission in his newspaper piece: "The truth is that, until I read your [Weintraub 1983] article, it never occurred to me that McKenzie's ingenious fixed-point construction was entirely

---

8 For a fuller discussion of the role of Abraham Wald in the story of the existence theorem for economic equilibrium, see Weintraub 1983.

independent from the Arrow-Debreu construction. I stand corrected. The closest thing I have to an excuse is the fact that I had four days in which to write the article" (Dorfman to Weintraub, June 27, 1984).

Arrow maintained his interest in Weintraub's historical reconstructions. In the commemorative *Cowles Fiftieth Anniversary* volume, Arrow (1991, 12) noted that for the general equilibrium work of the 1930s in Europe, "There is a good account of this period in Weintraub (1983; Hildreth 1986)." Much later, on September 19, 2010, he responded to a draft of Weintraub's (2011) paper on Lionel McKenzie. Apart from emphasizing priority in terms of his first technical report in December 1951, he also addressed the different approaches taken by McKenzie and by himself. Discussing the continuity of the demand correspondence that neither he nor Debreu got fully right in their independent drafts, he added: "I have always regarded this point as showing why our approach was better than McKenzie's, or Gale's, for that matter. Continuity of preferences is the primitive, not continuity of demand functions. The failure of continuity under certain circumstances is in fact an economically meaningful result."

He continued with comments about the appropriateness of the Kakutani theorem in comparison with the Eilenberg-Montgomery theorem.

> You [Weintraub] put great stress on Gérard's use of the Eilenberg-Montgomery theorem, rather than Kakutani's. I think you overdo the matter. Kakutani's theorem (like the original von Neumann fixed-point theorem, which is trivially equivalent to it) refers to convex sets. If I recall correctly (I haven't looked this up), Eilenberg and Montgomery postulate sets that are contractible to a point, a class of sets that includes convex sets. When one deals with convex sets, the two theorems are exactly the same, so that differentiating between them has little weight, in my judgment. (Arrow to Weintraub September 19, 2010)

Our own account of the proofs (see chapter 6) suggests that Debreu would not agree with Arrow on this point. The two theorems by Eilenberg and Kakutani are surely equivalent under certain conditions if the former is a generalization of the latter. Arrow thus downplays what Debreu emphasized as a difference between the Arrow-Debreu proof and McKenzie's. Arrow continued by distinguishing between the priority that was McKenzie's, that he and many others had long ago granted, and the influence of the Arrow and Debreu paper as it was determined by the larger community of economists.

> You seem to suggest that Arrow-Debreu have had the major credit. Whether or not that is true, I leave up to you, though frankly that would be my impression also. Yet I do not see how the tangled story of the publication process would have any effect on the audience, which is the determinant of credit and influence. As I say, as far as the world of economists (or even mathematical economists) was concerned, McKenzie had undisputed priority.

Arrow had no problem acknowledging McKenzie's priority. For Arrow, the proof was part of a larger project; he had no stake in any priority claim. Arrow believed that he, and Debreu, deserved credit not for publication priority but for the paper's influence. Debreu, though, as the following section suggests, had real difficulty identifying with *economists'* references to the Arrow-Debreu model.

## Debreu's 1983 Nobel Prize

A long process of lobbying and log-rolling among leading scientists precedes the announcement of any Nobel Prize. In economics, the award committee invites letters of nomination from not only leading figures in economics (past laureates, chairs of major departments around the world, etc.) but also authorities in the history of economics.[9] Kenneth Arrow was one of the key individuals who nominated

---

9 Nomination is by invitation only. In the case of the economics prize, the following have the right to submit nominations: members of the Royal Academy of Sciences, members of the Economics Prize Committee (five academicians plus ad hoc adjunct members), past recipients of the prize, permanent professors in relevant subjects at universities in Sweden, Denmark, Finland, Iceland, and Norway, holders of relevant chairs of at least six colleges and universities selected each year by the committee "with a view to insuring the appropriate distribution between different countries," and others from whom the academy solicits nominations. See "Nomination and Selection of Economic Sciences Laureates," *Nobelprize.org*, February 19, 2013, http://www.nobelprize.org/nobel_prizes/economics/nomination/index.html.

Gérard Debreu (KJAP, 20, Debreu). The Danish mathematician (and Debreu's friend) Karl Vind had asked Arrow, the general equilibrium macroeconomist Bent Hansen, and the Fields Medalist Steven Smale (both of the latter scholars from Berkeley) to write a joint letter of nomination. Vind had wanted to include as signatories a long list of others (Hahn, Meade, Malinvaud, Koopmans, Samuelson, Mas-Colell, and Sonnenschein), but Arrow and Hansen thought it better to keep the number small. Vind prepared the draft (ibid., October 6, 1982), Hansen rewrote it, and Smale and Arrow signed it. The differences between the nomination text and the later reasoning of the committee as expressed in the award statement are noteworthy. They prefigure the difficulty that Debreu was to have in accepting Nobel Prize credit on the terms given, terms remote from those he thought appropriate.

The nominators faced the difficulty that Debreu had hardly made an original contribution to economic theory but rather had transformed the mathematical standards of formulating economic theory. Thus Arrow, Hansen, and Smale referred to his indirect influence on many applied fields: "His achievements in pure theory are generally recognized by the economics professions and are emulated in many fields of applied theory such as public finance, capital theory, monetary economics, Marxist analysis of exploitation and class formation, and exhaustible resources just to mention a few" (KJAP, 20, Debreu). The mention of the Marxist analysis of exploitation referred to John Roemer (1980) and his general equilibrium approach to Marxian economics. This use of Debreu's work would not be mentioned again—to the contrary, his work was to be framed as anti-Marxist.

Another striking difference between the nomination and the award concerned the work the nominators singled out: not Debreu's paper with Arrow or his monograph the *Theory of Value* but instead his 1969 Irving Fisher lecture "Economies with a Finite Set of Equilibria" (1970). The nominators heralded a methodologically advanced paper that few economists could understand! Their letter recognized that, with respect to his many other contributions, an award honoring Debreu alone presented real difficulties: "[I]t is obviously difficult to decide who should be given credit for being the first to ask a given type of question or to give a new type of answer. There can, however, be no doubt that the present attitude—that it is a major weakness in a theory if there is no proof of existence—is highly influenced by Debreu's work." The nominators were clearly referring to McKenzie, Nikaido, and Gale but decided not to mention them in their letter. The Arrow, Hansen, and Smale letter must have convinced the Nobel award committee of Debreu's worthiness, but not on the terms for which the prize actually was given and for which the public would address him once he had been given the prize.

The way Debreu was informed about his winning the prize was a foretaste of what would follow in terms of the public perception of his achievement. On October 17, 1983, at quarter to four in the morning, a phone call awakened Debreu: "A radio station in New York City relayed the news from Stockholm and wanted to know what grade I would give President Reagan on his economic policy" (GDP, 14). After years of professional discreetness Debreu was now daily represented to the world as an economist, a member of a community in which he had always tried to keep a low profile. He who valued the anonymity of his work was now identified as the economist who "has mathematically proven the invisible hand." In receiving the Nobel it was as if his ivory tower was breached: the tower's moat that once provided protection had been filled in, paved over, and turned into a Hollywood Walk of Fame.

Speaking at his first press conference, Debreu assumed that he could accept the prize for the work he himself would have appreciated. With his opening remark he anticipated wrong-headed questions: "I do not want to discuss my views on the Reagan Administration's economic policies" (*New York Times*, October 18, 1983). He then made clear that he wanted to accept the prize as a mathematician, a stance quite confusing for his colleagues in the economics department.

> Unless you listened carefully, you wouldn't know that he got a prize in economics. There were almost no references to the economics department, no reference to the economics profession. He talked about the importance of funding the mathematical sciences, nothing about funding the economics profession. It was all about mathematics. (Steve Goldman, personal communication)

But after the press conference it must have dawned on him that even if he had wanted to accept the prize as a mathematician, it was not being given to him as a mathematician. This lesson awaited him when reading newspapers in the following days: "On many occasions during this strange period of seven months, the press was a source of a different kind of frustration in reflecting images of reality deeply impaired by, at times, distorting mirrors" (Debreu, in GDP, 14; translation by the authors). One of these deformations was an article in *Le Figaro Magazine* titled "The Superiority of Liberalism Is Mathematically Demonstrated" (October 3, 1984). This market liberal reference that caught the public's attention was not a fabrication by the press but was intrinsic to the Nobel committee's reasoning for granting Debreu the prize in the first place (ignoring Vind's reference to Roemer). The committee indeed suggested that Debreu and Adam Smith had the same object in mind, but Smith only had a vague intuition while Debreu had the scientific proof. This would turn out to be a splendid trope to explain his work to the public, and welcome as well for Marxists like Resnick and Wolff (1984, 30), who had always suspected the committee of political bias. The committee's own press release left no doubt about the connection between "their" Debreu and market liberalism:

> Adam Smith had already raised the question of how [self-interested market] decisions, apparently independent of one another, are coordinated. . . . [His answer was that] price systems automatically bring about the desired coordination of individual plans. Toward the end of the 19th century, Leon Walras formulated this idea in mathematical terms as a system of equations. . . . But it was not until long afterwards that this system of equations was scrutinized to ascertain whether it had an economically meaningful solution, i.e. whether this theoretical structure of vital importance for understanding the market system was logically consistent. Arrow and Debreu managed to prove the existence of equilibrium prices, i.e., they confirmed the internal logical consistency of Smith's and Walras' model of the market economy.

This politically conservative-sounding publicity challenged Debreu's persona of silence and discretion (Düppe 2012b). He did have the option of joining a political debate about the meaning of general equilibrium theory, but he decided not to respond.[10] In a letter to Siamack Shojai on May 20, 1987, he addressed his reticence with respect to misinterpretations of his work:

> I also have read the article by Resnick and Wolff in the *Monthly Review*, December 1984 without feeling [the need] to reply to the two authors. I have consistently tried to say clearly, sometimes with simple mathematics as in the enclosed paper, what the theory of general equilibrium had to contribute to the understanding of economic processes. I expect proponents of alternative theories to do the same, and to let readers judge. (GDP, 5, Shojai)

He wrote his Nobel lecture in this same frame of mind. He was aware that the reasons for giving him the prize would not match the terms in which he could explain his work. He tried to keep this dilemma hidden, thus avoiding any controversial interpretations of what he had done. We can see in his notes how he formulated the idea for his lecture: "Combination of non-trivial mathematics and interesting economics. . . . Focus on the mathematical examples that solved various economic problems: Pareto Optimum—Minkowski; Existence—Brower-Kakutani; Core—Lyapunov. Give precise history of the mathematics involved" (GDP, 8). Offering the public a "precise history of the mathematics involved" compelled Debreu to mention Lionel McKenzie's contribution:

> In addition to the work of Arrow and me, begun independently and completed jointly, Lionel McKenzie at Duke University proved the "Existence of an Equilibrium in Graham's Model of World Trade and Other Competitive Systems" (1954), also using Kakutani's theorem. A different approach taken independently by David Gale (1955) in Copenhagen, Hukukane Nikaido

10 His students did respond to the co-optation of their teacher into the neoliberal camp. In a letter to the editor of the *Wall Street Journal*, they "object[ed] to your editorial of October 20, which claimed Prof. Debreu proved 'the invisible hand of the market works. . . .' Your attempt to identify his theories with the particular economic policies you advocate does a disservice to both the man and the prize" (GDP, 6). Steve Smale, in the *California Monthly*, referred to Adam Smith but added that "the theory allows grossly unequal shares of the goods of society. Thus, substantial government mediation of the decentralized price system is required" (November 4, 1983).

> (1956) in Tokyo, and Debreu (1956) in Chicago permitted the substantial simplification given in my *Theory of Value* (1959) of the complex proof of Arrow and Debreu. (1984a, 270)[11]

The Royal Ceremony brought some of Debreu's ambivalence into the open. He understood that accepting credit might be as difficult as appearing to be worth crediting. Although letting oneself celebrate is not the same as living up to the reasons for being celebrated, being able to relate to the reasons in some way helps one enjoy the festivities. In the electric moment after the king of Sweden handed over the prize and the committee's reasoning was read aloud to the assemblage, Debreu gave his banquet speech. And he referred to the invisible hand! Surprisingly, he employed the idea to show how *little* he could relate to the consequences of his work that made him appear worthy of credit.

> [A] scientist knows that his motivations are often weakly related to the distant consequences of his work. The logical rigor, the generality, and the simplicity of his theories satisfy deep personal intellectual needs, and he frequently seeks them for their own sake. But here, as in Adam Smith's famous sentence, he seems to be "led by an invisible hand to promote an end which was not part of his intention," for his personal intellectual fulfillment contributes to promoting the social interest of the scientific community. . . . It was my great fortune to begin my career at a time when economic theory was entering a phase of intensive mathematization and when, as a result, the strength of that invisible hand had become irresistible. (Debreu 1984c)

Debreu's statement was masterful and sublimely ironic. He employed the metaphor of the invisible hand, but his use severed it from

11 The McKenzie papers include a copy of a mailgram sent on October 17, 1983, to Debreu: "Dear Gérard. Blanche and I are very happy that you have received your deserved award. Congratulations to you and our warm regards to you and to Francoise. Lionel." The response to McKenzie on November 2 was a form letter sent to several hundred people, which concludes, "To each of you I owe a personal note of thanks. Since an endless list of commitments will not permit me to write these notes, I shall remain totally in debt to all of you for the extraordinary support that you are giving me. Gérard Debreu."

any market—a fortiori neoliberal—interpretation. Debreu admitted that the primary concern of his intellectual life was not the "social interest of the scientific community." Instead he engaged in research for its own sake. He acknowledged that his influence had extended far beyond his own intentions. He had never considered the possible unintended consequences of mathematical virtues in economics. His banquet speech, read between the lines, was a way of saying that he was sorry for having caused a misunderstanding about the nature and import of mathematical economics.

In subsequent months Debreu had to learn how to be a public figure. At a dinner in honor of his Nobel at Berkeley he remarked, "After the trip to Stockholm, it becomes clear that one must reconsider the hypothesis that the world had a temporary fit of insanity, and that one has to come to terms with the new situation" (GDP, 14). His new fame created an interest in his work by those whom he had never considered to be his readers. Such individuals had not the slightest interest in the structure of economic theory but rather were concerned with the question that Debreu always carefully avoided: What Does That Mean? "The world's expectations didn't help either. Suddenly because he had done some remarkable work in economics my father was contacted by politicians, political activists, physicists and scientists in other disciplines and even by the Pope" (Chantal Debreu 2005). Political activists? The Pope? What could he have said apart from: "No, sorry, I did nothing that could help you in your mission, John Paul." Debreu was lost trying to find reasons why he should accept the Pope's invitation:

> I remember he called me to talk about it, and he said, "Should I go?" I said, "If you go, why are you going?" He didn't seem to be able to untangle [it]. There was all that honor associated with these famous powerful people calling him to meet him, to talk to him. But he didn't seem to have a sense of purpose any more. What's my role in this? (Chantal Debreu, personal communication)

In the years following his Nobel, Debreu increasingly broke away from the equilibrium of silence on which his life had previously been organized. Roger Hahn commented: "When he received his Nobel Prize, his personality changed. Of course, he was very proud, anybody would be proud. But he talked about it constantly, his stature, his responsibility, and his elevated position" (Roger Hahn, personal communication). Hildenbrand recalls his struggle with the public: He once said: "Koopmans was the only one who did it right: After three weeks he said, that's it, no more interviews, I am no more willing to appear as a Nobel laureate in public and thus could continue his work. [But] Debreu never again found his way back to regular work" (Werner Hildenbrand, personal communication).

It is perhaps too much to say that Debreu regretted accepting the prize. His ambivalence, however, was evident to his colleagues and family. In his reply to Arrow's AEA lunch speech in 1984, he spoke about the old "cross-discipline average age of laureates" and added, "These statistics might look bleak to future economics laureates, if it were not for the fact that the Nobel Medal has a reverse [effect], widely underestimated, and from which it may be wise to shield the younger members of our profession" (GDP, 14). In private, Debreu gathered information about everyone who had ever rejected the Nobel Prize. As his daughter recalled, he talked to a laureate in genetics who left academia after the prize. He gave clear voice to this sense of remorse at a Berkeley dinner in 1987:

> When that New York radio station called in the early morning of October, 17, 1983, I was not lucid enough to even think of making a cost-benefit analysis that accepting the prize should call for. When, some four hours later, the Secretary of the Swedish Academy of Sciences finally succeeded in reaching me on a constantly busy telephone, the question had answered itself in an obvious manner. (GDP, 14)

In those post-Nobel years some of his colleagues noted a personality change. His internal feelings of being apart, of being different that had been present since his orphaned childhood, were now accentuated. He and he alone was Berkeley's economics Nobel Prize winner. In his struggle to live up to his new identity, Debreu increasingly found it difficult to relate to his colleagues and students: "He began to think of himself as a public figure, rather than as a private individual" (Roger Hahn, personal communication). His wife put it simply: "after he was on the top, there was only one way he could go: and that was down. The Nobel was a catastrophe for him" (Françoise Debreu, personal communication). What a healing effect it would have had for Debreu, for the profession of economics, and for his family, if he had refused the prize! His daughter's words in her memorial to him were poignant:

> But from that day forward I believe he never felt he could live up to the honor that had been done him. His esteem for his own work did not match the high esteem that others had put upon it. . . . It was from that time onward that I saw my father withdraw from us. He was unwilling for any of us to see him as less than he had been judged in that brief shining moment in Stockholm. He could not live up to the myth that had been created around him. We deprived him and he deprived himself of his humanity, of his right to be flawed and his right to be loved no matter the current level of his achievements. (Chantal Debreu 2005)

# CONCLUSION

At age eighty-one Arrow commented that life philosophies, like economies, are never independent of history (1992, 43). In the previous chapters we explored this history of the professional careers of Kenneth Arrow, Gérard Debreu, and Lionel McKenzie as the discipline of economics changed in the post–World War II era. Neither their work nor their scientific selves were causal one for the other. Instead we have insisted that their own scientific identities, and the work that they did, mutually stabilized each other. Their scientific persona and their work became coherent together and can be best understood in relation one to the other.

Our narrative focused on issues of credit. Our protagonists' experiences of being granted, accepting, and failing to receive credit are consistent with traditional narratives constructed by historians of science. They have not so engaged historians of postwar economics (e.g., Ingrao and Israel 1990; Mirowski 2002; Amadae 2003). In contrast we believe that they are essential features of the emergent epistemic culture in economics during the postwar period; thus before closing we want to step back and revisit these themes. One point of reference will be Robert Merton's (1973) first important discussion of scientific credit and priority, one that has been developed in many different ways over the years by succeeding generations of sociologists of science. The other point of reference is our repersonalization of the work credited and our discussions of the scientific personae. Merton's original analysis will help us reframe our preceding remarks on life-writing in science: credit and the creation of the scientific self are not unrelated.

## Credit and the Scientific Self

The careers of three gifted men stand for the events that changed economics in the second half of the twentieth century century. A discipline that had been salient in various political and moral discourses became a closed field with a distinct and clearly demarcated body of knowledge. Although there were many themes and subthemes associated with that change in economics, we selected the proofs of the existence of equilibrium for models of a competitive economy as representative. The use of formal models and mathematical proof techniques in economic theory fit well with the idea that in all sciences conclusions had to be demonstrably derived from simpler assumptions. As the "door that opens into the house of analysis," as Mas-Colell later called it (Mas-Colell, Whinston, and Green 1995, 584), the proof was vital for establishing the norms that one has to meet even today in order to call oneself an economist.

As that proof was published in two simultaneously written papers involving three authors, our choice allowed us to unfold historically, in a highly personalized manner, a scientific work that on the surface represents itself as an impersonal expression. The existence proof represents an intellectual culture that was no longer rooted in various traditions going back to founders and their visions. Lost was "Keynes's vision" or "Schumpeter's view of capitalism." They were replaced by discussions of topics like "forecasting models of the U.S. economy," "rational expectations and supply shocks," and "models for derivative pricing." The three different careers of our protagonists stand for three different scientific characters, each consistent with economists' acceptance of economic knowledge as impersonal knowledge. Our emphasis on the impersonal character of knowledge, in other words, allows us to unfold a complex, rather than banal, relationship between the life of the scientist and the products of science. Insofar as anonymity characterizes the changing epistemic culture in economics, the site where this change can be translated into a narrative is the scientist's persona. Thus the adaptation to a regime of anonymous knowledge is historically manifest in a changing relationship between science and the scientist. The contingency and absence of the person in a regime of anonymous knowledge compel the construction of an historical narrative to deal with this dimension of science.

The problem of scientific credit became pertinent as the motif of this repersonalization of anonymous knowledge. Public credit and anonymous knowledge are intertwined. In contrast to literary and artistic communities, credit in science is not given for the innovation of new styles or interpretations that are naturally identified with the author or artist but for the novelty of discovery and methods that do not themselves bear the names of its discoverers. You can't sign a theorem. Originality in art is associated with a personalized body of work that usually bears the mark of the artist, even if it as abstract as Barnett Newman or John Cage. Personal expressions of knowledge witness the experience of developing a personal style of thinking (think, for example, of the travel accounts of nineteenth-century zoologists and their description of "getting to know" something). As the historian of science Mario Biagioli put it: "A scientific claim is not rewarded as the material inscription of the scientist's personal expression but as a nonsubjective statement about nature" (2003, 254). Alternatively, as Debreu said about his work: "Even though a mathematical economist may write a great deal, it usually remains impossible to make, from his works, a reliable conjecture about his personality" (2001, 4). And so the decision about who earns credit is not trivial. *Assigning credit is complex because it brings about, rather than responds to, the public identification of the scientist and the scientific product.* Statements of truth in modern science are an anonymous product of human creativity, and so credit is necessary if the scientist is to appear in public at all. The *surfacing* of the person in discussions of crediting the scientist bares the underside of the production of modern knowledge.

We are not historians, or hagiographers, giving or denying credit. Our project instead has examined the negotiations of credit in the economy of science. Scientific credit is that which supplements the attachment of persons to scientific achievements in a regime of impersonal, mathematical knowledge. Credit is in fact nothing but the way subjects are made the historical origins of ideas. When historicizing credit we thus historicize the notion of the scientific self—vis-à-vis changing social structures—as the origin of ideas.

To be sure, scientific credit has always been contested in intellectual regimes. Yet it is its hidden character in a regime of impersonal knowledge that makes it particularly salient for the history of science. It informs us about both the plurality of ways such a culture shapes a scientific self and the problems one encounters in constructing such a cohesive self in science. Three different scientific selves—for each of them, how did issues of credit inflect their paths?

## Arrow's Communitarianism

The early 1950s proof of the existence of competitive equilibrium in a market economy by McKenzie and by Arrow and Debreu, the process of writing we unraveled in chapter 6, seems to present a case of essentially simultaneous discovery, presentation, and publication. This comes as no surprise given what we, in accord with Arrow, have said about the intellectual atmosphere at Cowles in the early 1950s. The existence proof was "in the air" at least after the 1949 activity analysis conference and the concomitant revival of interest in von Neumann's use of a fixed-point theorem. A new sort of mathematical sensibility was to be necessary for any proof attempt. The 1949 air was, as we have seen, scented with convexity, separating hyperplanes, and fixed-point theory. Only the few at Cowles had learned to breathe this air.

The Cowles Commission was one of the first research communities in economics that was neither based on apprenticeship nor held together by a scholarly leader surrounded by a group of colleagues and students (e.g., "Hayekians" or "Marshallians" or "Keynesians").[1] The Cowles Commission stands for a communal effort loosely organized around shared and uninterrogated standards of techniques, with different and even conflicting visions regarding the meaning of economic science. At the time of the activity analysis conference in 1949 it was unclear whether mathematical reasoning in economic

1 Randall Collins has argued that intellectual networks associated with such charismatic intellectual figures shape the intellectual content of their ideas: "In a very strong sense, networks are the actors on the intellectual stage. . . . The network dynamics of intellectual communities provides an internal sociology of ideas" (1998, xviii). Such an approach to studying the emergence of the new postwar economics, although potentially powerful, requires prior historical studies such as our own.

theory would turn out to be a subfield of mathematics (as part of its research agenda, namely programming, indeed did) or if it would mainly serve to toughen the methodological standards of economic research.

Arrow was an ideal member of the Cowles community. Cowles made him think of economic research as being "more and more of a cooperative matter, requiring teams of individuals trained along similar lines" (KJAP, Accession 2008–0037). He had an eclectic sensitivity to the variety of problems that could be approached with technical verve, and that left him rather indifferent to the sorts of issues of priority that he had faced earlier with Duncan Black:

> Here was Gérard who had exactly the same thing without me. McKenzie had almost exactly the same thing, not quite as good, but pretty much the same thing. Here we had at least two fine scholars with essentially the same idea and, I think, that if neither of them had existed, someone else would have appeared on the scene. Of course, the von Neumann-Nash tradition had created tools. Once the tools are in place, somebody is bound to pick them up. . . . If I had not done it, somebody else would have. . . . Incidentally, both of us were influenced by Koopmans's formulation of production (activity analysis), that was another common ground, in this respect for the formulation, not the proofs. As far as consumer theory is concerned, we both started from the same point; it is all Hicks (and Hicks himself is not all that new in this respect). (Arrow, in Feiwel 1987, 195–96)

Arrow had not considered issues of priority at the time of the existence proof not only because he was uninformed about McKenzie's paper but also because for him the existence proof was an integral part of a larger project pursued by the entire Cowles community. It was, after all, Arrow who suggested a joint collaboration with Debreu for the sake of producing a better result than they each could have produced on their own. For Arrow, proving existence was a well-defined project in a community of scholars who crossed institutional boundaries between Cowles, RAND, and the mathematics department, and disciplinary boundaries between mathematics, statistics, engineering, and social science. One can easily imagine that Arrow, had he heard of McKenzie's work at an early stage, would have proposed joint authorship to both Debreu and McKenzie.

Arrow did not seem to strategize regarding future credit for the existence proof. This is not to ascribe a superior moral intelligence to Arrow. Rather Arrow represented a regime of truth production whose social norms were those of shared values. An individual's endowment of specific capacities—personal genius and charisma—mattered less than did his openness to others' new ideas. Arrow stands for a new form of scientific author who functions only in a community that is united less by a philosophical commitment to truth than by a strong sense of purpose. For example, in 1968 Arrow published jointly with four other scholars using the pseudonym of "Archen Minsol"—a composition of Kenneth Arrow, Hollis B. Chenery, Bagicha S. Minhas, and Robert M. Solow (Minsol 1968; RMSP, 53, C1).

Merton called the scientific value that Arrow internalized the "communism" of knowledge (1973, 273–75). He might have said about Arrow what Joseph Priestley had said about John Aubrey (as quoted by Merton): "whenever he discussed a new fact in science, he instantly proclaimed it to the world, in order that other minds might be employed upon it besides his own" (Merton 1973, 317). If this communal character of knowledge production becomes the norm, the fact of coauthorship and the fact of simultaneous discovery are indeed the normal case in the production of knowledge. As Merton framed the matter:

> But I should like now to develop the hypothesis that, far from being odd or curious or remarkable, a pattern of independent multiple discoveries in science is in principle the dominant pattern rather than a subsidiary one. It is the singletons—discoveries made only once in the history of science—that are the residual cases, requiring special explanation. Put even more sharply, the hypothesis states that all scientific discoveries are in principle multiples, including those that . . . on the surface appear to be singletons. (1973, 356)

As a sociologist writing before the science studies movement reshaped the sociology of scientific knowledge, Merton treated scientific values as having been imposed by social forces. If these shared scientific values are the condition of scientific production, there cannot be any "singletons" that are not hidden "multiples": "The sheer fact that multiple discoveries are made by scientists working independently of one another testifies to the further crucial fact that . . . they are responding to much the same social and intellectual forces that impinge upon them all. In a word, the Robinson Crusoe of science is just as much a figment as the Robinson Crusoe of old-fashioned economics" (Merton 1973, 375).

And yet even if we accept Merton's hypothesis, the problem of the simultaneity of scientific discoveries for the scientist is that credit is not given to entire communities but to individuals. In contrast to sociologist Merton, we as historians of science believe it is equally important to see how older regimes of knowledge are still present in an officially communitarian regime. These dual layers produce the kind of negotiations that we have discussed in detail in the preceding chapters. Credit, specifically public credit for science, is an atavism. It represents an older surviving image of scientific practice that does not take place in professionalized communities but in the mind and/or laboratory of the solitary scientist.[2] Truth, in that older regime, is produced by individuals with higher qualities, such as more powerful intellects, who are in a particular state of mind free of mundane concerns and free from imposed social constraints that corrupt the search for truth. There the pursuit of truth is deemed to be a self-less and self-constrained practice that grants the scientist moral distinctness. Universal truth requires self-less recipients. And so credit must first be individual and second "symbolic," uncorrupted by unclean incentives like money (see Biagioli 2003, 254ff).

The Nobel Prize clearly fashions itself in the image of this old regime of knowledge in which universal knowledge is attached to

2 Scientific solitude is a vast subject in the history of science that still waits to be fully explored. For one excellent historical account, see Shapin 1990, and for notes on its cultural role in post–World War II U.S. society, see Hollinger 1996, 164–66.

individuals by means of symbolic value.[3] In Stockholm scientific credit is the reward, rather than the incentive, for scientists who selflessly sacrifice their life for the solitary pursuit of truth without being questioned about their mundane existence.[4] The achievements of a laureate are celebrated in public as the result not of a specific organizational structure but of specific qualities of individuals: intelligence, brightness, genius, work habits, and so forth, that is, in terms of a sensibility or receptiveness to truth. And this receptiveness requires one to be alienated from oneself. It requires self-abnegation with respect to one's socially engaged daily life for the sake of some higher goal. It requires one's self-constraint to achieve moral distinction. The publically proclaimed assignment of credit speaks in hagiographic language to accord with this representation of knowledge. It was this kind of credit that Gérard Debreu was eager to receive.

## Debreu between Norms

In our narrative Debreu appears to have had the most complex personal investments about receiving credit. His strategizing with respect to credit was also the subtlest. He sought credit for creating impersonal knowledge that would allow him to be recognized even as he sustained his self-imposed emotional half-life. His insistence on rigor and purity on the one hand and his strong need for personal credit on the other reinforced one another: the more anonymous his work, the greater his need to become attached to his work in other than immediate ways. Debreu, given his self-protective character and his unwillingness to be identified with a vision or purpose of science, invested all his personal ambitions into the generality and impersonality of his work. Yet it was he who hesitated to accept the collaboration with Arrow unless he was able to secure those parts of the proof that he wanted to claim as his own; it was he who kept

---

3 The Nobel Prize of course does come with a career-changing money payment. In Nobel Prize winner Robert Lucas's divorce agreement with his first wife, for instance, there was an active court-approved clause concerning the joint division of any future Nobel Prize monies that he might receive.

4 That is why when James Watson wrote, in his *Double Helix* (1968), about his and Francis Crick's racing Linus Pauling for the Nobel-worthy solution to the structure of DNA, he transgressed against the new regimes by revealing that the old one still lived.

secret from McKenzie his work-in-progress. Debreu did not embrace Arrow's communal vision of applied science. He strategized in order to protect himself and to appear worth crediting at the same time. He replayed the conflicts of his childhood.

Within the regime of epistemic virtues that Debreu represents there is indeed a close link between anonymous knowledge and the strategizing and anxiety related to priority fights. Only anonymous truth can be separated from the scientific persona. Only such truths can be discovered rather than created. Only that which can be discovered can be discovered first, and only what can be discovered first can be cause for scientific reward.

The tragedy of Debreu's career is that although he finally received credit, he was not given credit on his own terms. What is credited in public is not the person who happens to discover a communally relevant result; what is credited is the set of qualities of a person that allows him to discover that relevant result. In our account, Debreu's role in constructing the existence proofs of the early 1950s exemplifies the tension between personal ambition in science and impersonal knowledge in science. A scientist's interest in credit supplements that scientist's need for expressive content. Since for a mathematician, particularly a Bourbakist, such an expressive need is systematically frustrated, the significance of reward is multiplied. Merton referred to a similar personal conflict when speaking of priority issues that emerge from the tension between two contradictory values of science, *originality* and *humility*:

> The components of this ambivalence [toward priority] are fairly clear. After all, to insist on [one's] originality by claiming priority is not exactly humble and to dismiss one's priority by ignoring it is not exactly to affirm the value of originality. As a result of this conflict, scientists come to despise themselves for wanting that which the institutional values of science have led them to want. (1973, 305)

Debreu gave us a clear sense of this self-loathing in his sometimes subtle, sometimes open attempts to claim both priority and superiority of his proof. We repeatedly documented how his emotional struggle, hidden from public view, was ever-present in both his coauthorship and the peer-review process. But we agree with Merton that strategizing for credit is not to be dismissed as mere self-interested behavior:

> To say that these frequent conflicts over priority are rooted in the egotism of human nature, then, explains next to nothing; to say that they are rooted in the contentious personalities of those recruited by science may explain part, but not enough; to say, however, that these conflicts are largely a consequence of the institutional norms of science itself comes closer, I think, to the truth. For, as I shall suggest, it is these norms that exert pressure upon scientists to assert their claims, and this goes far toward explaining the seeming paradox that even those meek and unaggressive men, ordinarily slow to press their own claims and other spheres of life, will often do so in their scientific work. (1973, 293)

We have not made a case, nor do we believe one can be made, for the notion that economic scientists reveal any specific character strengths or flaws as preconditions for priority fights to take place. We know of no psychological framework that would predict such disputes or even provide a framework for interpreting them. Instead the preceding narrative exhibited Arrow's and Debreu's different behaviors as they negotiated the same institutions of science. This suggests that such emotional struggles are associated with individuals' different personalities rather than the products of social forces as Merton argues.[5]

## McKenzie and the Matthew Effect

As we assessed the problem of credit from McKenzie's point of view, we injected a third consideration associated with a regime of impersonal knowledge production. McKenzie did sense what was "in the air." However, he believed that others would not easily credit

---

5 There is, however, an older, and essentially psychoanalytic, literature that grew from the writings of Freud and Jung and that sought to locate the source of literary and scientific production in the unconscious of the writer or scientist. Kubie argues, for instance, that "the life of a young scientist challenges our educational system from top to bottom with a series of unsolved problems. . . . Among these are certain subtle problems, arising out of unrecognized neurotic forces, which are basically important both in the choice and in the pursuit of scientific research as a career" (1953, 596). Merton, aware of such arguments, found them unhelpful.

those working in rather remote outposts of the academic world. Having been failed by Hicks's supervision and the Oxford examiners (Roy Harrod and A. M. Henderson),[6] McKenzie was excluded from breathing the new air. McKenzie knew from the 1952 Chicago meeting at the latest that he needed to do two things: first, he had to show that his international trade model was as general as Arrow-Debreu's, and second, that he could assert priority. The controversial question of whether one model was better or equivalent to the other made competing priority claims inevitable. Though he was in part successful in convincing some members of the community on both matters, he never received the credit equal to that given to Arrow and Debreu. McKenzie was acutely aware of his professional marginalization. If Arrow personified the communal character of knowledge production and Debreu represented the conflict between older and newer norms of knowledge production, then McKenzie reminds us of the economics community's *institutional hierarchies*.

Our own explanation for McKenzie's receiving less credit than did Arrow and Debreu is thus based on what Robert Merton (1973, 445, reprinted from 1968) called the "Matthew effect." He noted that there is a tendency to give greater credit to those who are better known and quoted the Gospel according to St. Matthew: "For unto every one that hath shall be given, he shall have abundance: but from him that hath not shall be taken away even that which he hath." In 1954 McKenzie, lacking a Ph.D., was teaching at his alma mater Duke University, a segregated Southern university with a weak graduate program and no presence in economic theory. He had been floundering at Duke in the late 1940s; as he wrote to his friend Ian Little in 1950: "The purpose of the trip to Chicago is to get some fresh air, intellectually speaking, learn a bit of statistics, and perhaps make some kind of new start. I don't know." When McKenzie received his doctorate in 1957 from Princeton (by collecting his published papers and submitting them as a thesis), he moved from Duke to the University of Rochester, which had no doctoral program in economics at that time. McKenzie was very active in enlarging his community by

6 In a letter to Weintraub (January 6, 1982), McKenzie identified his examiners as Harrod and A. M. Henderson. Henderson, then at Manchester, was the external examiner.

supervising Ph.D. students. His community was built around his own narrative of the events during the early 1950s. His students internalized the traumatic event of his having been denied proper credit. In the postwar regime of truth, McKenzie represents the undervalued and underestimated scholar.

In contrast, Arrow in the mid-1950s had been a Ph.D. student connected to Abraham Wald and Harold Hotelling, had been at Cowles and RAND, and was teaching at Stanford. Debreu had been a student of mathematicians like Henri Cartan at the École Normale Supérieure and, in Chicago, talked regularly with mathematicians like Saunders Mac Lane, John Milnor, I. N. Herstein, Morton Slater, and André Weil. Taken up after the war by Maurice Allais and "discovered" in Salzburg by Robert Solow, Debreu came to the United States to the Cowles Foundation, moved with that group to Yale, and eventually settled at the University of California at Berkeley. In the post-1954 years, both Arrow and Debreu worked in highly prestigious academic institutions, while McKenzie remained by comparison an academic outsider.

The Matthew effect and the anonymous character of scientific knowledge are related. Given that the idea of proving existence for a competitive market model had been in the air, it is not surprising that the proof had been pursued simultaneously, and independently, using similar proof techniques. At the time there was no essentialist reason to state that one proof was going to be better than another. If knowledge is impersonal then similar results appear to be equivalent. In this way *mathematical knowledge, as rigid it might be, creates space for social factors to intrude on the community process of assigning credit*. In examining McKenzie's career we thus illustrated a third aspect of the culture of anonymous knowledge, namely that social factors, contrary to what one might otherwise assume, are increasingly important for determining credit. If the community, as we observed for Arrow's case, becomes the agent of science, not being in this community is an even stronger reason not to be credited. The nature of mathematical expression hardly ever compels an economist's allegiance to one form of mathematical knowledge over another. If one economic theorist's proof uses measure theory, another's uses nonstandard analysis, while a third's uses differential geometry, the community that can barely distinguish one mathematical proof technique from another gives credit on social not mathematical grounds.

In the final referee report on the Arrow-Debreu paper prior to publication in *Econometrica* associate editor Nicholas Georgescu-Roegen wrote, "I have the highest opinion of the authors and I trust Debreu's mathematics" (Weintraub and Gayer 2001, 434). As we have seen in chapter 6, McKenzie was not similarly trusted and certainly not by his last *Econometrica* referee, Gérard Debreu.

# CODA

> I haven't done collaborative work on my historical projects [except for my two articles with Ted Gayer], but I'm about to coauthor a book on the history of existence of general equilibrium with a young German researcher Till Düppe. He and I have been corresponding back and forth because he's unearthed, and has access to, the Debreu papers, [and he has] lengthy interviews with members of the Debreu extended family. I'd been planning to make my McKenzie material into a book, because I have so much of it, and I started thinking, "God, I wish I had the Debreu papers." I wrote to him asking whether he was the least bit interested in my McKenzie material. He jumped at it! And so we are going to write the book together. It should be interesting, because it's my first attempt at joint work on a [large] historical project. As you get older, sometimes you want to try something different. Can we bring this off? I don't know. It may just crash and burn, but at the moment it seems like fun.
>
> **WEINTRAUB IN BOWMAKER 2012, 433,**
> **FROM AN INTERVIEW, JANUARY 14, 2011**

Coauthorship was an important theme in the preceding historical study. One element of our story was the making of the Arrow-Debreu model that later appeared to have emerged out of one mind, that of Arrow-Debreu. But this study was not only *about* coauthorship but a book written in coauthorship. More than a simple coincidence, at least at one point this fact intervened in our text when we reflected upon, and distanced ourselves from, the earlier involvement of the older author, E. Roy Weintraub, in chapter 8 on the history of credit to our protagonists. The reader might thus be interested to learn about some relevant aspects of coauthoring the present study as these echo some of the book's concerns.

There are parallels between what we have said about the collaboration of Arrow and Debreu and our own collaboration. Like Arrow and Debreu, we met for the first time in person at a point when a large portion of the text was already written. And just as the different cultural socializations played out with Arrow and Debreu, so too did they play out with us: as the one learned of the flat hierarchies of U.S. university professors, the other learned of a continental inclination to quote obscure continental philosophers. Before Arrow and Debreu joined forces, they each had independently produced their own partial proofs. We, too, had our own accounts of the Arrow-Debreu collaboration (Düppe 2012a) and the chronology of McKenzie's proof (Weintraub 2011). Unlike Koopmans's acting as the Arrow-Debreu matchmaker, though, modern technology served that function as we learned of each other's work via an online streaming of Weintraub's talk on September 20, 2010, in a workshop in São Paulo, Brazil, organized by Pedro Duarte. We then gave credit to one another in our respective articles. But in our accounts we did not treat the question of credit symmetrically. While Düppe played it down by showing "that Arrow and Debreu did not share the same interest, . . . played different roles during the process of their work, and drew different lessons from it" (Düppe 2012a, 492), the first anonymous referee of the book manuscript felt compelled to warn the editor that Weintraub "seems to be on something of a crusade to give McKenzie what he believes is his just due in the history of economics." And so, in a three-year process of letting the text grow through mutual edits, the book emerged as our joint version, the result of a struggle that for Düppe meant an emancipation from treating history as an illustration of philosophy and for Weintraub required a commitment to bring his decades of necessarily incomplete narratives to closure.

Issues of credit, however, did not intrude in the collaboration. What is salient is rather the intergenerational character of this collaboration: seventy may be the new fifty, but it is not the new thirty. The age and reputation difference between us did not inform the collaboration. It was not a senior-junior coauthorship where the junior does the actual work, nor one in which the senior author assigns the junior author minor and tiresome writing tasks. The collaboration turned out, unlike that of Arrow and Debreu, to be nonstrategic: ideas, structures, rhetorical strategies, illustrations, and individual sentences were joint decisions.

There are yet more essential differences between our collaboration and that of Arrow and Debreu, differences that bring us back to the main theme of our book: the impersonality of mathematical expressions. Coauthorship in that epistemic regime, as we have seen for the case of Arrow and Debreu, forced them to make concessions concerning their own interests and, more important, it increased the stakes with respect to scientific credit. In our own case it is a category mistake to ask what of this book can be credited to whom: *as historians we never produce anonymous knowledge*. The history of science is always a personal intellectual endeavor in which the author is inseparable from the narrative. As a consequence, even as we started from separate texts, hardly any line of this book now can be assigned to one or the other author. Weintraub has developed a distinct style in his writings but it became clear at an early stage that this book would be different. In one of his messages Weintraub wrote to Düppe: "It's gotten to the point where I can no longer hear my voice, or your voice, but I am beginning to hear 'our voice' in the narrative. Frankly, it's a most remarkable process!" (January 10, 2012). Coauthorship in the history of economics raises questions about creativity rather than credit.

There is one dimension of our text, however, that might appear to add an impersonal dimension to the history of science: archival work. If we "discover" and "unearth" letters in archives, can we then claim priority? Does crediting historical work like ours entail a judgment regarding priority of archival discovery? Would everyone who goes to the archives we visited produce an identical book? The answer to all three questions is clearly "No." Simply finding stuff is mindless. Archival *research* is personal. It requires interpretation, and context, and thus a creative imagination. In fact, the archival material that we used in chapter 5 has already been employed by another historian of economics (Backhouse 2012) for an account rather different from ours without there being any question of priority. By no means thus does the study at hand discourage further work using even the same archival material. Other historians will pose different questions and find different answers in the archives. The preceding study used these documents in order to reconstruct both the changing body and image of economic knowledge through the lives of those who brought about this change. Other historians will see different potentialities when contextualizing both the archive-texts and the knowledge that the archives are constructed to inform. Indeed the historiography of economic science is a record of historians' representations and re-representations, interpretations and reinterpretations. In this way the past remains alive.

# ACKNOWLEDGMENTS

At earlier stages of this project, both Kenneth Arrow and Lionel McKenzie graciously offered their support for, and commentary on portions of, our joint project. Weintraub's interviews with Debreu in 1992 facilitated his early work on this material. Düppe's understanding of Debreu benefited greatly from conversations with Robert M. Anderson, Truman Bewley, Marcel Boiteux, Monique Florenzano, Gaël Giraud, Steve Goldman, Roger Hahn, Werner Hildenbrand, Roy Radner, Herbert Scarf, and Martin Shubik, as well as family members David Arrow and Françoise Debreu. Weintraub's understanding of McKenzie's role was further clarified in conversations with Ron Jones, Roman Pancs, and Stan Engerman in Rochester, and with McKenzie's former Duke student Craufurd Goodwin and his Rochester students Randi Novak and Jerry Green. Düppe is particularly grateful for the most valuable support and encouragement of Debreu's daughter Chantal Debreu.

Early versions of parts of this book were presented at the University of São Paulo, Duke University, the University of Cergy-Pontoise Paris, St. Petersburg State University, the University of Mainz, the University of Athens, Humboldt University Berlin, the University of British Columbia Vancouver, and the University of Manchester. The authors received helpful comments from many individuals at those seminars and workshops, particularly Ivan Boldyrev, Beatrice Cherrier, Pedro Duarte, Philippe Fontaine, Yann Giraud, Verena Halsmayer, Bruna Ingrao, M. Ali Khan, Robert Leonard, Harro Maas, Tiago Mata, and Nikolaus Wolf. At the authors' joint presentation at Humboldt University in Berlin, Lorraine Daston and Werner Hildenbrand were immensely helpful in signaling where some revisions of the arguments might be useful. At Duke's CHOPE Workshop, comments from Neil De Marchi, Craufurd Goodwin, Kevin Hoover, and Bruce Caldwell improved the narrative line. Roger Backhouse, Mary Morgan, and anonymous readers for Princeton University Press read the entire manuscript. We are extremely grateful for the care and attention they gave to our work and especially for their encouragement.

Several portions of this book are drawn from articles written for different audiences. Düppe's (2012a, 2012b) helped shape some material in chapters 3 and 6, and Weintraub's (2011) informed some of the narrative in chapters 2 and 6. We thank the editors of the *Journal of the History of Economic Thought*, the *History of Political Economy*, and the *Journal of Economic Perspectives* for their permission to use revised portions of those papers.

Superb transcriptions of Weintraub's recorded drafts were made by Maxine Borjon. Our editor at Princeton University Press, Seth Ditchik, was firm in his belief in this project. Beth Clevenger at the Press ran interference for us when we got bogged down in details and rescued us from many errors. She has the rare gift of finding ways to say "yes" to authors. She also gave us a real gift of copy editor Jennifer Backer, whose sense of language and good judgment were greatly appreciated. We thank them all.

The staff of Duke's Rubinstein Rare Book and Manuscript Library, particularly Bob Byrd, Will Hansen, and Elizabeth Dunn, were our partners as we worked in the Economists Papers Projects collections of the papers of Arrow, McKenzie, Samuelson, Solow, Georgescu-Roegen, Hoover, and de Vyver. Our project could never have been contemplated, let alone developed and completed, without the superb collection development work of Byrd and Hansen. Düppe also received help from the Bancroft Library at the University of California, Berkeley, in guiding him through the Debreu papers, as well as from the Yale University Library.

# REFERENCES


Abella, Alex. 2008. *Soldiers of Reason: The RAND Corporation and the Rise of the American Empire*. Orlando: Harcourt.

Abram, Morris B. 1982. *The Day Is Short: An Autobiography*. New York: Harcourt Brace Jovanovich.

Albers, Donald J., et al. 1986. "An Interview with George B. Dantzig: The Father of Linear Programming." *College Mathematics Journal* 17(4): 293–314.

Aldrich, John. 2003. "William H. Riker." Pp. 312–24 in *Encyclopedia of Public Choice*, ed. Charles Rowley and Friedrich Schneider. New York: Kluwer Academic Publishers.

Amadae, Sonja M. 2003. *Rationalizing Capitalist Democracy: The Cold War Origins of Rational Choice Liberalism*. Chicago: University of Chicago Press.

Anderson, Robert. 2005. Gérard Debreu Memorial Speech. University of California, Berkeley, March 4.

Arrow, Kenneth J. 1949. "On the Use of Winds in Flight Planning." *Journal of Meteorology* 6: 150–59.

———. 1951a. "Alternative Proof of the Substitution Theorem of Leontief Models in the General Case." Pp. 155–64 in *Activity Analysis of Production and Allocation*, ed. Tjalling Koopmans. Cowles Commission Monograph 13. New York: John Wiley.

———. 1951b. *Social Choice and Individual Values*. New York: Wiley.

———. 1951c. "An Extension of the Basic Theorems of Classical Welfare Economics." Pp. 507–32 in *Proceedings of the Second Berkeley Symposium on Mathematical Statistics and Probability*, ed. Jerzy Neyman. Berkeley: University of California Press.

———. 1951d. "Mathematical Models in the Social Sciences." Pp. 129–54 in *Policy Sciences in the United States*, ed. Harold D. Lasswell and Daniel T. Lerner. Stanford: Stanford University Press.

———. 1952. "Le rôle des valeurs boursières pour la répartition la meilleure des risques." Pp. 1–8 in *International Colloquium on Econometrics*. Paris: CNRS.

———. 1962. "The Economic Implications of Learning by Doing." *Review of Economic Studies* 29: 155–73.

———. 1963. "Uncertainty and the Welfare Economics of Medical Care." *American Economics Review* 53 (3): 941–73.

———. 1991. "Cowles in the History of Economic Thought." Pp. 1–24 in *Cowles Fiftieth Anniversary: Four Essays and an Index of Publications*, ed. Alvin K. Klevorick. New Haven: The Cowles Foundation.

———. 1992. "I Know a Hawk from a Handsaw." Pp. 42–50 in *Eminent Economists: Their Life Philosophies*, ed. Michael Szenberg. Cambridge: Cambridge University Press.

———. 2005. Gérard Debreu Memorial Speech. University of California, Berkeley, March 4.

———. 2008. "George Dantzig in the Development of Economic Analysis." *Discrete Optimization* 5: 159–67.

———. 2009. "Some Developments in Economic Theory since 1940: An Eyewitness Account." *Annual Review of Economics* 1: 1–16.

Arrow, Kenneth J., and Gérard Debreu. 1954. "Existence of an Equilibrium for a Competitive Economy." *Econometrica* 22 (3): 265–90.

———, eds. 2001. *Landmark Papers in General Equilibrium Theory, Social Choice and Welfare: Edward Elgar Companion to General Equilibrium*. Cheltenham: Edward Elgar.

Arrow, Kenneth, and Alain Enthoven. 1956. "A Theorem on Expectations and the Stability of Equilibrium." *Econometrica* 24 (3): 288–93.

Arrow, Kenneth, and Frank Hahn. 1971. *General Competitive Analysis*. Amsterdam: Elsevier.

Arrow, Kenneth, and Leonid Hurwicz. 1958. "On the Stability of Competitive Equilibrium I." *Econometrica* 26: 522–52.

Arrow, Kenneth, Henry D. Block, and Leonid Hurwicz. 1959. "On the Stability of the Competitive Equilibrium II." *Econometrica* 27 (1): 82–109.

Arrow, Kenneth J., Samuel Karlin, and Patrick Suppes, eds. 1960. *Mathematical Methods in the Social Sciences: Proceedings of the First Stanford Symposium, June 24, 1959*. Stanford: Stanford University Press.

Asprey, William. 1990. *John von Neumann and the Origins of Modern Computing*. Cambridge, MA: MIT Press.

Aubin, David. 1997. "The Withering Immortality of Nicholas Bourbaki: A Cultural Connector at the Confluence of Mathematics, Structuralism, and the Oulipo in France." *Science in Context* 10 (2): 297–342.

Augier, Mie, and James G. March. 2011. *The Roots, Rituals, and Rhetorics of Change: North American Business Schools after the Second World War*. Stanford: Stanford University Press.

Backhouse, Roger. 1998. "The Transformation of U.S. Economics, 1920–1960: Viewed through a Survey of Journal Articles." *History of Political Economy* 30 (Supplement): 85–107.

———. 2012. "Paul Samuelson, RAND and the Cowles Commission Activity Analysis Conference, 1947–1949." Working paper.

Backhouse, Roger, and Mauro Boianovsky. 2013. *Transforming Modern Macroeconomics: Exploring Disequilibrium Microfoundations, 1956–2003.* Cambridge: Cambridge University Press.

Backhouse, Roger, and Philippe Fontaine, eds. 2010. *The History of the Social Sciences since 1945.* Cambridge: Cambridge University Press.

Bagioli, Mario. 2003. "Rights or Rewards? Changing Frameworks of Scientific Authorship." In *Scientific Authorship: Credit and Intellectual Property in Science,* ed. Mario Biagioli and Peter Galison. New York: Routledge.

Begle, Edward G. 1950. "A Fixed Point Theorem." *Annals of Mathematics* 51 (3): 544–50.

Bergson, Abram. 1938. "A Reformulation of Certain Aspects of Welfare Economics." *Quarterly Journal of Economics* 52 (2): 310–34.

Biagioli, Mario, and Peter Galison, eds. 2003. *Scientific Authorship: Credit and Intellectual Property in Science.* New York: Routledge.

Bini, Piero, and Luigino Bruni. 1998. "Intervista a Gérard Debreu." *Storia del Pensiero Economico* 35: 3–29.

Black, Duncan. 1948. "On the Rationale of Group Decision-making." *Journal of Political Economy* 56 (1): 23–34.

Blank, Rebecca M. 1991. "The Effects of Double-Blind versus Single-Blind Reviewing: Experimental Evidence from the *American Economic Review.*" *American Economic Review* 81 (5): 1041–67.

Blaug, Mark. 1980. *The Methodology of Economics: Or How Economists Explain.* Cambridge: Cambridge University Press.

Bockman, Johanna. 2011. *Markets in the Name of Socialism: The Left-Wing Origins of Neoliberalism.* Palo Alto, CA: Stanford University Press.

Bockman, Johanna, and Michael A. Bernstein. 2008. "Scientific Community in a Divided World: Economists, Planning, and Research Priority during the Cold War." *Comparative Studies in Society and History* 50 (3): 581–613.

Boumans, Marcel. 2012. "Observations in a Hostile Environment: Morgenstern on the Accuracy of Economic Observations." Pp. 110–31 in *History of Observations in Economics,* ed. Harro Maas and Mary S. Morgan. Durham: Duke University Press.

Bourbaki, Nicolas. 1949. "Foundations of Mathematics for the Working Mathematician." *Journal of Symbolic Logic* 14 (1): 1–8.

———. 1950. "The Architecture of Mathematics." *American Mathematical Monthly* 57 (4): 221–32.

———. [1939] 1968. *Elements of Mathematics: Theory of Sets.* Reading, MA: Addison-Wesley.

Bowmaker, Simon W. 2012. *The Art and Practice of Economics Research: Lessons from Leading Minds.* Northampton, MA: Edward Elgar.

Breit, William, and Roger W. Spencer. 1995. *Lives of the Laureates*. Cambridge, MA: MIT Press.

Bruner, Jerome S. 1960. *The Process of Education*. Cambridge, MA: Harvard University Press.

Buck, Paul H., et al. 1945. *General Education in a Free Society*. Cambridge, MA: Harvard University Press.

Burns, Arthur F., and Wesley C. Mitchell. 1946. *Measuring Business Cycles*. New York: National Bureau of Economic Research.

Bush, Vannevar. 1945. *Science, the Endless Frontier: A Report to the President*. Washington, DC: Office of Scientific Research and Development.

Caldwell, Bruce J. 2004. *Hayek's Challenge*. Chicago: University of Chicago Press.

Chadarevian, Soraya de, and Nick Hopwood, eds. 2004. *Models: The Third Dimension of Science*. Stanford: Stanford University Press.

Coats, Alfred William, ed. 1996. *The Post-1945 Internationalization of Economics*. History of Political Economy Annual Supplement 28 (5). Durham: Duke University Press.

Collins, Harry, and Martin Kusch. 1998. *The Shape of Actions: What Humans and Machines Can Do*. Cambridge, MA: MIT Press.

Collins, Randall. 1998. *The Sociology of Philosophies: A Global Theory of Intellectual Change*. Cambridge, MA: Belknap Harvard.

Conant, James B. 1947. *On Understanding Science*. New Haven: Yale University Press.

Corry, Leo. 1989. "Linearity and Reflexivity in the Growth of Mathematical Knowledge." *Science in Context* 3 (2): 409–40.

———. 1992. "Nicolas Bourbaki and the Concept of Mathematical Structure." *Synthese* 92: 315–48.

Cottle, Richard W. 2010. "A Brief History of the International Symposia on Mathematical Programming." *Mathematical Programming Series B* (125): 207–33.

Cowles Foundation. 1951. "Cowles Report 1951." http://cowles.econ.yale.edu/P/reports/index.htm.

Crane, Diana. 1972. *Invisible Colleges: Diffusion of Knowledge in Scientific Communities*. Chicago: University of Chicago Press.

Davis, George H. 1990. "A Quarter Century of Water Resources Research." *Water Resources Research* 26 (1): 2–4.

Dasgupta, Dipankar. 2010. "Memoirs: Is Life Worth Living? It Depends Upon the Liver!" Boloji.com.

Debreu, Chantal. 2005. Gérard Debreu Memorial Speech. University of California, Berkeley, March 4.

Debreu, Gérard. 1949. "Les fins du système économique." *Revue d'Economie Politique* 600–615.

———. 1951a. "The Coefficient of Resource Utilization." *Econometrica* 19 (3): 273–92.
———. 1951b. "Saddle Point Existence Theorems." *Cowles Commission Discussion Paper Mathematics* 412.
———. 1952a. "An Economic Equilibrium Existence Theorem." *Cowles Commission Discussion Paper Economics* 2032.
———. 1952b. "A Social Equilibrium Existence Theorem." *Proceedings of the National Academy of Sciences* 38 (8): 886–93.
———. 1953. "The Continuity of Multivalued Functions." *Cowles Commission Discussion Paper Economics* 2079.
———. 1954. "A Classical Tax-Subsidy Problem." *Econometrica* 22 (1): 14–22.
———. 1956. "Market Equilibrium." *Proceedings of the National Academy of Sciences* 42 (11): 876–78.
———. 1959. *Theory of Value: An Axiomatic Analysis of Economic Equilibrium.* New York: Wiley.
———. 1960a. "On 'An Identity in Arithmetic.'" *Proceedings of the American Mathematical Society* 11 (2): 220–21.
———. 1960b. "Une économiques de l'incertain." *Economie Applique* 13 (1): 111–16.
———. 1962. "New Concepts and Techniques for Equilibrium Analysis." *International Economic Review* 3 (3): 257–73.
———. 1970. "Economies with a Finite Set of Equilibria." *Econometrica* 38 (3): 387–92.
———. 1974. "Excess Demand Functions." *Journal of Mathematical Economics* 1: 15–21.
———. 1983a. *Mathematical Economics: Twenty Papers of Gérard Debreu.* Econometric Society Monographs. Cambridge: Cambridge University Press.
———. 1983b. "Mathematical Economics at Cowles." In *Cowles Fiftieth Anniversary Volume*, ed. Alvin K. Klevorick. New Haven: The Cowles Foundation.
———. 1984a. "Economic Theory in the Mathematical Mode." *American Economic Review* 74 (3): 267–78.
———. 1984b. "Autobiography." In *Les Prix Nobel: The Nobel Prizes 1983*, ed. Wilhelm Odelberg. Stockholm: Nobel Foundation.
———. 1984c. "Banquet Speech." In *Les Prix Nobel: The Nobel Prizes 1983*, ed. Wilhelm Odelberg. Stockholm: Nobel Foundation.
———. 1987. "Die stillen Stars." TV interview with Frank Elstner, August 18, 1986. Mainz: ZDF.
———. 1991. "Random Walk and Life Philosophy." *American Economist* 35 (2): 3–7.
———. 1998. "Foreword: Economics in a Mathematics Colloquium." Pp. 1–4 in *Ergebnisse eines Mathematischen Kolloquiums*, ed. Karl Menger, Egbert Dierker, and Karl Siegmund. Wien: Springer.

———. 2001. "Introduction." Pp. 1–4 in *Economics Essays: A Festschrift for Werner Hildenbrand*, ed. Gérard Debreu, Wilhelm Neuefeind, and Walter Trockel. Berlin: Springer.

Debreu, Gérard, and Israel N. Herstein. 1953. "Nonnegative Square Matrices." *Econometrica* 21 (4): 597–607.

Debreu, Gérard, and Herbert Scarf. 1963. "A Limit Theorem on the Core of an Economy." *International Economic Review* 4 (3): 235–46.

Dickson, David. 1988. *The New Politics of Science*. Chicago: University of Chicago Press.

Dieudonné, Jean. 1970. "The Work of Nicolas Bourbaki." *American Mathematical Monthly* 77 (2): 134–45.

Dorfman, Robert. 1983. "A Nobel Quest for the Invisible Hand." *New York Times*, October 23, pp. 15–16.

———. 1984. "The Discovery of Linear Programming." *Annals of the History of Computing* 6 (3): 283–95.

Dorfman, Robert, Paul Samuelson, and Robert Solow. 1958. *Linear Programming and Economic Analysis*. New York: McGraw-Hill.

Dorman, Joseph. 1998 [DVD 2005]. *Arguing the World: The New York Intellectuals in Their Own Words*. USA, PBS. 109 minutes.

———. 2001. *Arguing the World: The New York Intellectuals in Their Own Words*. Chicago: University of Chicago Press.

Duffie, Darrell, and Hugo Sonnenschein. 1989. "Arrow and General Equilibrium Theory." *Journal of Economic Literature* 27 (2): 565–98.

Düppe, Till. 2009. "Listening to the Music of Reason: Nicolas Bourbaki and the Phenomenology of the Mathematical Experience." Working paper.

———. 2012a. "Arrow and Debreu De-homogenized." *Journal of the History of Economic Thought* 34 (4): 491–551.

———. 2012b. "Gérard Debreu's Secrecy: His Life in Order and Silence." *History of Political Economy* 44 (3): 413–49.

Düppe, T., and E. R. Weintraub. 2014. "Siting the New Economic Science: The Cowles Commission's Activity Analysis Conference of June 1949." *Science in Context* 27 (3).

Edgeworth, Francis Y. 1881. *Mathematical Psychics: An Essay on the Application of Mathematics to the Moral Sciences*. London: Kegan Paul.

Eilenberg, Samuel, and Deane Montgomery. 1946. "Fixed Point Theorems for Multi-valued Transformations." *American Journal of Math-ematics* 68 (2): 214–22.

Emmett, Ross B. 2011. "Sharpening Tools in the Workshop: The Workshop System and the Chicago School's Success." Pp. 93–115 in *Building Chicago Economics: New Perspective on the History of America's Most Powerful Economics Program*, ed. Robert van Horn, Philip Mirowski, and Thomas Stapleford. Cambridge: Cambridge University Press.

Epstein, Roy J. 1987. *A History of Econometrics*. Amsterdam: North-Holland.
Erickson, Paul, et al. 2013. *How Reason Almost Lost Its Mind: The Strange Career of Cold War Rationality*. Chicago: University of Chicago Press.
———. 2014. "The World the Game Theorists Made." Unpublished manuscript.
Feiwel, George, ed. 1987. *Arrow and the Ascent of Modern Economic Theory*. New York: New York University Press.
Feuer, Lewis S. 1982. "The Stages in the Social History of Jewish Professors in American Colleges and Universities." *American Jewish History* 71 (4): 432–65.
Finetti, Bruno de. 1949. "Sulle stratificazione convesse." *Annali I Matematica: Pura ed applicata*, ser. 4, 30 (1): 173–83.
Fish, Stanley E. 1980. *Is There a Text in This Class: The Authority of Interpretive Communities*. Cambridge, MA: Harvard University Press.
Fleck, Ludwig. [1935] 1979. *Genesis and Development of a Scientific Fact*. Chicago: University of Chicago Press.
Fourcade, Marion. 2010. *Economists and Societies: Discipline and Profession in the United States, Britain, and France, 1890s to 1990s*. Princeton: Princeton University Press.
Friedman, Milton, and Leonard J. Savage. 1948. "Utility Analysis of Choices Involving Risk." *Journal of Political Economy* 56 (4): 279–304.
Gale, David. 1955. "The Law of Supply and Demand." *Mathematica Scandinavica* 3: 33–44.
———. 1956. "Mathematics and Economic Models." *American Scientist* 44 (1): 33–44.
———. 1960. *The Theory of Linear Economic Models*. New York: McGraw-Hill.
———. 1963. "A Note on Global Instability of Competitive Equilibrium." *Naval Research Logistics Quarterly* 10: 81–87.
Gale, David, and Lloyd S. Shapley. 1962. "College Admissions and the Stability of Marriage." *American Mathematical Monthly* 69 (1): 9–15.
Gallagher, Noel. 2005. "Gérard Debreu Dies at 83: First of Four Berkeley Economists to Win Nobel Prize over 18-Year Span." *UC Berkeley Public Affairs* 8 (1): 1.
Goethe, Johann W. [1836] 1998. *Conversations of Goethe with Johann Peter Eckermann*. London: Da Capo Press.
Gordon, Robert A. 1976. "Rigor and Relevance in a Changing Institutional Setting." *American Economic Review* 66 (1): 1–14.
Gray, Jeremy. 2008. *Plato's Ghost: The Modernist Transformation of Mathematics*. Princeton: Princeton University Press.
Gross, Alan G. 1998. "Do Disputes over Priority Tell Us Anything about Science?" *Science in Context* 11: 161–79.
Guedj, Denis. 1985. "Nicolas Bourbaki, Collective Mathematician: An Interview with Claude Chevalley." *Mathematical Intelligencer* 7 (2): 18–22.

Hankins, Thomas L. 2007. "Biography and the Reward System in Science." Pp. 93–104 in *The History and Poetics of Scientific Biography*, ed. Thomas Söderqvist. Burlington, VT: Ashgate.

Hardy, Godfrey H. [1908] 1944. *A Course in Pure Mathematics*. Cambridge: Cambridge University Press.

———. [1940] 1969. *A Mathematician's Apology*. Cambridge: Cambridge University Press.

Hartley, Leslie P. 1954. *The Go-Between*. New York: Knopf.

Heims, Steve J. 1980. *John Von Neumann and Norbert Wiener: From Mathematics to the Technologies of Life and Death*. Cambridge, MA: MIT Press.

Henderson, David. 2010. "Memorial to Lionel McKenzie." http://econlog.econlib.org/archives/2010/10/lionel_mckenzie.html.

Henderson, James M., and Richard E. Quandt. 1958. *Microeconomic Theory: A Mathematical Approach*. New York: McGraw-Hill.

Herstein, Israel N., and John Milnor. 1953. "An Axiomatic Approach to Measurable Utility." *Econometrica* 21 (2): 291–97.

Hicks, John R. 1939. *Value and Capital*. Oxford: Oxford University Press.

———. 1961. "Prices and the Turnpike. I.: The Story of a Mare's Nest." *Review of Economic Studies* 28 (2): 77–88.

———. 1965. *Capital and Growth*. Oxford: Oxford University Press.

Hildreth, Clifford. 1986. *The Cowles Commission in Chicago, 1939–1955*. Berlin: Springer.

Hollinger, David A. 1996. *Science, Jews, and Secular Culture: Studies in Mid-Twentieth Century American Intellectual History*. Princeton: Princeton University Press.

Horn, Karen I. 2009. *Roads to Wisdom: Conversations with Ten Nobel Laureates in Economics*. Cheltenham: Edward Elgar.

Horn, Rob v., et al., eds. 2011. *Building Chicago Economics: New Perspectives on the History of America's Most Powerful Economics Program*. Cambridge: Cambridge University Press.

Hotelling, Harold. 1938. "The General Welfare in Relation to Problems of Taxation and of Railway and Utility Rates." *Econometrica* 6 (3): 242–69.

Hurwicz, Leonid. 1961. "Review of *Theory of Value*." *American Economic Review* 51 (3): 414–17.

Hutchison, Terence W. 1977. *Knowledge and Ignorance in Economics*. Oxford: Basil Blackwell.

Ikeo, Aiko. 2006. *Nihon no Keizaigaku* [Economics in Japan: The Internationalization of Economics in the Twentieth Century]. Nagoya: Nagoya University Press. Translated by author.

———. 2009. "How Modern Algebra Was Used in Economic Science in the 1950s." Presented at the History of Economics Society Annual Meeting, June 27–29, 2009, Denver.

Ingrao, Bruna, and Georgio Israel. 1990. *The Invisible Hand: Economic Equilibrium in the History of Science*. Cambridge, MA: MIT Press.
Israel, Georgio, and Ana M. Gasca. 2009. *The World as a Mathematical Game: John von Neumann and Twentieth Century Science*. Basel: Birkhäuser.
Jaffé, William. 1967. "Walras's Theory of Tâtonnement: Critique of Recent Interpretations." *Journal of Political Economy* 75 (1): 1–19.
Jones, Ronald W. 2012. "Lionel McKenzie: A Recruit's View of the Early Rochester Days." *International Journal of Economic Theory* 8 (1): 9–12.
Kakutani, Shizuo. 1941. "A Generalization of Brouwer's Fixed Point Theorem." *Duke Mathematical Journal* 8 (3): 457–59.
Karabel, Jerome. 2005. *The Chosen*. New York: Houghton Mifflin.
Kelly, Jerry S. 1987. "An Interview with Kenneth J. Arrow." *Social Choice and Welfare* 4: 43–62.
Kevles, Daniel J. 1996. "The National Science Foundation and the Debate over Postwar Research Policy, 1942–45: A Political Interpretation of *Science: The Endless Frontier*." *Isis* 68 (1): 4–26.
Kindleberger, Charles P. 1991. *The Life of an Economist: An Autobiography*. Cambridge, MA: Basil Blackwell.
Klein, Lawrence. 1947. *The Keynesian Revolution*. London: Macmillan.
———. 1991. "Econometric Contributions of the Cowles Commission, 1944–47." *Banca Nazionale del Lavoro Quarterly Review* 177: 107–17.
Knorr-Cetina, Karin. 1991. *Epistemic Cultures: How the Sciences Make Knowledge*. Cambridge, MA: Harvard University Press.
Koopmans, Tjalling C. 1947. "Measurement without Theory." *Review of Economic Statistics* 29 (3): 161–72.
———, ed. 1951a. *Activity Analysis of Production and Allocation: Proceedings of a Conference*. New York: Wiley.
———. 1951b. "Efficient Allocation of Resources." *Econometrica* 19 (4): 455–65.
———. 1957. *Three Essays on the State of Economic Science*. New York: McGraw-Hill.
Krueger, Alan B. 2003. "An Interview with Edmond Malinvaud." *Journal of Economic Perspectives* 17 (1): 181–98.
Kubie, Lawrence S. 1953. "Some Unsolved Problems of the Scientific Career." *American Scientist* 41 (4): 596–613.
Kuhn, Harold W. 2008. "57 Years of Close Encounters with George." Kuhn on Dantzig.pdf. from http://www2.informs.org/History/dantzig/articles_kuhn.html.
Kuhn, Thomas S. [1962] 1996. *The Structure of Scientific Revolutions*. Chicago: University of Chicago Press.
Lange, Oskar. 1942. "The Foundations of Welfare Economics." *Econometrica* 10 (3/4): 215–28.
Latour, B. 1987. *Science in Action*. Cambridge, MA: Harvard University Press.

Leggon, Cheryl B. 2001. *The Scientist as Academic: The American Academic Profession*. New Brunswick, NJ: Transaction Publishers.

Leonard, Robert. 2010. *Von Neumann, Morgenstern, and the Creation of Game Theory: From Chess to Social Science, 1900–1960*. Cambridge: Cambridge University Press.

Lester, Richard A. 1950. "In Memoriam: Frank Dunstone Graham, 1890–1949." *American Economic Review (Papers and Proceedings)* 40 (2): 585–87.

Lipset, Seymour M., and J. Everett Carll Ladd. 1971. "Jewish Academics in the United States: Their Achievements, Culture, and Politics." *American Jewish Yearbook* 72: 89–128.

Livingston, Eric. 1999. "Cultures of Proving." *Social Studies of Science* 29 (6): 867–88.

Louçã, Francisco. 2007. *The Years of High Econometrics: A Short History of the Generation That Reinvented Economics*. New York: Routledge.

Louçã, Francisco, and Sofia Terlica. 2011. "The Fellowship of Econometrics: Selection and Diverging Views in the Province of Mathematical Economics, from the 1930s to the 1950s." *History of Political Economy* 43 (Supplement): 57–85.

Lowen, Rebecca S. 1997. *Creating the Cold War University: The Transformation of Stanford*. Berkeley: University of California Press.

Lyman, Stanford M. 1994. "A Haven for Homeless Intellects: The New School and Its Exile Faculties." *International Journal of Politics, Culture, and Society* 7 (3): 493–512.

Machlup, Fritz. 1952. "Introductory Remarks." *American Economic Review* 42 (2). Papers and Proceedings of the Sixty-fourth Annual Meeting of the American Economic Association.

Mackavey, William R., Janet E. Malley, and Abigail J. Stewart. 1991. "Remembering Autobiographically Consequential Experiences: Content Analysis of Psychologists' Accounts of Their Lives." *Psychology and Aging* 6 (1): 50–59.

Mackay, Charles. [1852] 1980. *Extraordinary Popular Delusions and the Madness of Crowds*. New York: Harmony Books.

Mac Lane, Saunders. 1989a. "Mathematics at the University of Chicago: A Century of Mathematics in America." Pp. 127–54 in *American Mathematical Society History of Mathematics, Volume 2: A Century of Mathematics in America II*, ed. Peter L. Duran. Providence, RI: American Mathematical Society.

———. 1989b. "Topology and Logic at Princeton." In *A Century of Mathematics in America. Part II*. P. L. Duran, 217–22. Providence RI: American Mathematical Society.

Macrae, Norman. 1992. *John von Neumann*. New York: Pantheon Books.

Mantel, Rolf. 1974. "On the Characterization of Aggregate Excess Demand." *Journal of Economic Theory* 7: 348–53.

Mas-Colell, Andrew. 1974. "An Equilibrium Existence Theorem without Complete or Transitive Preferences." *Journal of Mathematical Economics* 1 (3): 237–46.

Mas-Colell, Andrew, Michael Whinston, and Jerry Green. 1995. *Microeconomic Theory*. Oxford: Oxford University Press.

Mashaal, Maurice. 2006. *Bourbaki: A Secret Society of Mathematicians*. Providence, RI: American Mathematical Society.

McKenzie, Lionel W. 1951. "Ideal Output and the Interdependence of Firms." *Economic Journal* 61 (244): 785–803.

———. 1954a. "On Equilibrium in Graham's Model of World Trade and Other Competitive Systems." *Econometrica* 22 (2): 147–61.

———. 1954b. "Specialization and Efficiency in World Production." *Review of Economic Studies* 21 (3): 165–80.

———. 1955a. "Equality of Factor Prices in World Trade." *Econometrica* 23 (3): 239–57.

———. 1955b. "Specialization in Production and the Production Possibility Locus." *Review of Economic Studies* 23 (1): 56–64.

———. 1955c. "Competitive Equilibrium with Dependent Consumer Preferences." Pp. 277–94 in *Proceedings of the Second Symposium in Linear Programming*, ed. H. A. Antosiewicz. Washington, DC: National Bureau of Standards and Directorate of Management Analysis.

———. 1957. "Demand Theory without a Utility Index." *Review of Economic Studies* 24 (3): 185–89.

———. 1959. "On the Existence of General Equilibrium for a Competitive Market." *Econometrica* 27 (1): 54–71.

———. 1960. "Stability of Equilibrium and the Value of Positive Excess Demand." *Econometrica* 28 (3): 606–17.

———. 1967. "The Inversion of Cost Functions: A Counter-Example." *International Economic Review* 8 (3): 271–78.

———. 1981. "The Classical Theorem on Existence of Competitive Equilibrium." *Econometrica* 49 (3): 819–41.

———. 1999. "A Scholar's Progress." *Keio Economic Studies* 36 (1): 1–12.

———. 2012. "The Early Years of the Rochester Department of Economics." *International Journal of Economic Theory* 8 (2): 227–35.

Medewar, Peter B. 1968. "Lucky Jim: A Review of *The Double Helix* by James D. Watson." *New York Review of Books*, March 28.

Merton, Robert K. 1957. "Priorities in Scientific Discovery: A Chapter in the Sociology of Science." *American Sociological Review* 22 (6): 635–59.

———. 1963. "Resistance to the Systematic Study of Multiple Discoveries in Science." *European Journal of Sociology* 4: 250–82.

———. 1973. *The Sociology of Science*. Chicago: University of Chicago Press.

Minsol, Archen. 1968. "Some Tests of the International Comparisons of Factor Efficiency with the CES Production Function: A Reply." *Review of Economics and Statistics* 50 (4): 477–79.

Mirowski, Philip. 2002. *Machine Dreams: Economics Becomes a Cyborg Science*. Cambridge: Cambridge University Press.

Mirowski, Philip, and E. Roy Weintraub. 1994. "The Pure and the Applied:

Bourbakism Comes to Mathematical Economics." *Science in Context* 7 (2): 245–72.

Mitra, Tapan, and Kazuo Nishimura, eds. 2009. *Equilibrium, Trade, and Growth: Selected Papers by Lionel W. McKenzie*. Cambridge, MA: MIT Press.

Morgan, Mary S. 2012. *The World in the Model: How Economists Work and Think*. Cambridge: Cambridge University Press.

Morgan, Mary S., and M. Masterman, eds. 1999. *Models as Mediators: Perspectives on Natural and Social Science*. Cambridge: Cambridge University Press.

Morgan, Mary S., and Malcolm Rutherford, eds. 1999. *From Interwar Pluralism to Postwar Neoclassicism*. History of Political Economy Annual Supplement 30 (5). Durham: Duke University Press.

Morgenstern, Oskar. 1941. "Professor Hicks on Value and Capital." *Journal of Political Economy* 49 (3): 361–93.

Nash, John F. 1950. "Equilibrium Points in n-Person Games." *Proceedings of the National Academy of Sciences of the United States of America* 36 (1): 48–49.

Neumann, John von. 1928. "Zur Theorie der Gesellschaftspiele." *Mathematische Annalen* 100: 295–320. Trans. 1959. "On the Theory of Games of Strategy." Pp. 13–42 in *Contributions to the Theory of Games IV*, ed. Robert D. Luce and Albert W. Tucker. Princeton: Princeton University Press.

———. 1937. "Über ein ökonomisches Gleichungssystem und eine Verallgemeinerung des Brouwerschen Fixpunksatzes." Pp. 73–83 in *Ergebnisse eines Mathematischen Kolloquiums 1935–36*, vol. 8, ed. Karl Menger. Leipzig: Deuticke. Trans. George Morton. 1945. "A Model of General Economic Equilibrium." *Review of Economic Studies* 13(1): 1–9.

———. 1947. "The Mathematician." *Works of the Mind* 1 (1): 180–96.

Neumann, John von, and Oskar Morgenstern. [1944] 1953. *Theory of Games and Economic Behavior*. Princeton: Princeton University Press.

Nikaido, Hukukane. 1956. "On the Classical Multilateral Exchange Problem." *Metroeconomica* 8: 135–45.

Oehler, Kay. 1990. "Speaking Axiomatically: Citation Patterns to Early Articles in General Equilibrium Theory." *History of Political Economy* 22 (1): 101–12.

Patinkin, Don. 1956. *Money, Interest and Prices*. New York: Harper and Row.

Peleg, Bezalel, and Menahem E. Yaari. 1970. "Markets with Countably Many Commodities." *International Economic Review* 11 (3): 369–77.

Pias, Claus, ed. 2003/2004. *Cybernetics—Kybernetic. The Macy-Conferences 1946–1953, Band 1: Transactions/Protokoll, Band 2: Documents/Dokumente*. Zürich: Diaphenes.

Porter, Theodore M. 1996. *Trust in Numbers: The Pursuit of Objectivity in Science and Public Life*. Princeton: Princeton University Press.

Price, Reynolds. 2009. *Ardent Spirits: Leaving Home, Coming Back*. New York: Scribner.

"Report of the Chicago Meeting, December 27–29, 1952." 1953. *Econometrica* 21 (3): 463–90.

Resnick, Stephen A., and Richard D. Wolff. 1984. "The 1983 Nobel Prize in

Economics: Neoclassical Economics and Marxism." *Monthly Review* (December): 29–46.
"Rhodes Scholar Appointed Professor of Economics." 1957. *Rochester Review* 18 (5): 4–5.
Riker, William H., and Peter C. Ordeshook. 1973. *An Introduction to Positive Political Theory*. Englewood Cliffs, NJ: Prentice-Hall.
Rizvi, Abu Turab S. 2006. "The Sonnenschein-Mantel-Debreu Results after 30 Years." *History of Political Economy* 38: 228–45.
Robinson, Joan. 1951. *The Pure Theory of International Trade: Collected Economic Papers*. New York: Augustus M. Kelley.
Roemer, John. 1980. "A General Equilibrium Approach to Marxian Economics." *Econometrica* 48 (2): 505–30.
Russell, Bertrand. 1919. *Introduction to Mathematical Philosophy*. London: George Allen & Unwin.
Samuelson, Paul A. 1947. *Foundations of Economic Analysis*. Cambridge, MA: Harvard University Press.
———. 1952. "Economic Theory and Mathematics: An Appraisal." *American Economic Review* 42 (2): 56–66.
———. 1954. "Some Psychological Aspects of Mathematics and Economics." *Review of Economics and Statistics* 36 (4): 380–86.
———. 2009. "Three Moles." *Bulletin of the American Academy* (Winter): 83–84.
Sapolsky, Harvey M. 1990. *Science and the Navy: The History of the Office of Naval Research*. Princeton: Princeton University Press.
Scarf, Herbert. 1960. "Some Examples of Global Instability of Competitive Equilibrium." *International Economic Review* 1 (3): 157–72.
———. 1991. "The Origins of Fixed Point Methods." Pp. 126–34 in *History of Mathematical Programming: A Collection of Personal Reminiscences*, ed. Jan K. Lenstra, Alexander Rinnooy Kan, and Alexander Schrijver. Amsterdam: North-Holland.
———. 1995. "Tjalling Charles Koopmans (August 28, 1910–February 26, 1985)." In *Biographical Memoirs* 67: 263–91.
Scherer, Frederic M. 2000. "The Emigration of German-Speaking Economists after 1933." *Journal of Economic Literature* 38 (3): 614–26.
Schlesinger, Karl. [1933] 1935. "Über die Produktionsgleichungen der ökonomischen Wertlehre." Pp. 1–6 in *Ergebnisse eines mathematischen Kolloquiums 1933–34*, vol. 6, ed. Karl Menger. Leipzig: Deuticke.
Shackle, George L. S. 1967. *The Years of High Theory: Invention and Tradition in Economic Thought, 1926–1939*. Cambridge: Cambridge University Press.
Shafer, Wayne, and Hugo Sonnenschein. 1975. "Some Theorems on the Existence of Competitive Equilibrium." *Journal of Economic Theory* 11 (1): 83–93.
Shapin, Steven. 1990. "'The Mind Is Its Own Place': Science and Solitude in Seventeenth-Century England." *Science in Context* 4 (1): 191–218.
———. 2010. *Never Pure: Historical Studies of Science as if It Was Made by*

*People with Bodies, Situated in Space, Time, and Society, and Struggling for Credibility and Authority.* Baltimore: Johns Hopkins University Press.

Shapin, Steven, and Simon Schaffer. 1985. *Leviathan and the Air-Pump: Hobbes, Boyle, and the Experimental Life.* Princeton: Princeton University Press.

Shubik, Martin. 1959. "Edgeworth Market Games." Pp. 267–78 in *Contributions to the Theory of Games IV*, ed. Robert D. Luce and Albert W. Tucker. Princeton: Princeton University Press.

———. 1961. "Review of *The Theory of Value.*" *Canadian Journal of Economics and Political Science* 27 (1): 133.

———. 1977. "Competitive and Controlled Price Economies: The Arrow-Debreu Model Revisited." Pp. 213–24 in *Equilibrium and Disequilibrium in Economic Theory*, ed. Gerhard Schwödiauer. Dordrecht: D. Reidel.

———. 1992. "Game Theory at Princeton, 1949–1955: A Personal Reminiscence." *History of Political Economy* 24 (Supplement): 151–63.

Slater, Morton L. 1950. "Lagrange Multipliers Revisited." *Cowles Commission Discussion Paper Mathematics* 403.

Smale, Steve. 1974. "Global Analysis and Economics II." *Journal of Mathematical Economics* 1: 1–14.

Söderqvist, Thomas. 1997. "Existential Projects and Existential Choice in Science: Science Biography as an Edifying Genre." Pp. 45–84 in *Telling Lives: Studies of Scientific Biography*, ed. Michael Shortland and Richard Yeo. Cambridge: Cambridge University Press.

———. 2003. *Science as Autobiography: The Troubled Life of Niels Jerne.* New Haven: Yale University Press.

Solovey, Mark. 2013. *Shaky Foundations: The Politics-Patronage-Social Science Nexus in Cold War America.* New Brunswick, NJ: Rutgers University Press.

Sonnenschein, Hugo. 1972. "Market Excess Demand Functions." *Econometrica* 40 (3): 549–63.

Starr, Ross M. 2008. "Arrow, Kenneth Joseph (born 1921)." *The New Palgrave Dictionary of Economics*, ed. Steven N. Durlauf and Lawrence E. Blume. New York: Palgrave Macmillan.

Stone, Marshall H. 1989. "Reminiscences of Mathematics at Chicago: A Century of Mathematics in America." Pp. 183–90 in *A Century of Mathematics in America, Part II*, ed. Peter Duren with the assistance of Richard A. Askey and Uta C. Merzbach. Providence, RI: American Mathematical Society.

Tarski, Alfred. 1941. *Introduction to Logic and to the Methodology of Deductive Sciences.* New York: Oxford University Press.

Varian, Hal R. 1978. *Microeconomic Analysis.* New York: Norton.

Volterra, Vito. 1906. "Les mathématiques dans les sciences biologiques et sociales." *La Revue du Mois* 1: 1–20.

Wald, Abraham. 1934. "Über die eindeutige positive Lösbarkeit neuen Produktionsgleichungen I." Pp. 12–18 in *Ergebnisse eines mathematischen Kolloquiums, 1933–34*, vol. 6, ed. Karl Menger. Leipzig: Franz Deuticke.

———. 1935. "Über die Produktionsgleichungen der ökonomischen Weltlehre (Mitteilung II)." Pp. 1–6 in *Ergebnisse eines mathematischen Kolloquiums, 1934–35*, vol. 7, ed. Karl Menger. Leipzig: Franz Deuticke.

———. [1936] 1951. "Über einige Gleichungssysteme der mathematischen Ökonomie." *Zeitschrift für Nationalökonomie* 7 (5): 637–70. Trans. 1951. "On Some Systems of Equations of Mathematical Economics." *Econometrica* 19 (4): 368–403.

Walker, Donald A. 1988. "Ten Major Problems in the Study of the History of Economic Thought." *History of Economics Society Bulletin* 10: 99–115.

Ward, Benjamin. 1972. *What's Wrong with Economics*? New York: Basic Books.

Warwick, Andrew. 2003. *Masters of Theory*. Chicago: University of Chicago Press.

Watson, James D. 1968. *The Double Helix: A Personal Account of the Discovery of the Structure of DNA*. New York: Atheneum.

Weil, André. 1991. *The Apprenticeship of a Mathematician*. Boston: Birkhäuser.

Weintraub, E. Roy. 1977. "Microfoundations of Macroeconomics: A Critical Survey." *Journal of Economic Literature* 15 (1): 1–23.

———. 1979. *Microfoundations: The Compatibility of Microeconomics and Macroeconomics*. Cambridge: Cambridge University Press.

———. 1983. "The Existence of a Competitive Equilibrium: 1930–1954." *Journal of Economic Literature* 21 (1): 1–39.

———. 1985. *General Equilibrium Analysis: Studies in Appraisal*. New York: Cambridge University Press.

———. 1987. "Stability Theory via Liapunov's Method: A Note on the Contribution of Takuma Yasui." *History of Political Economy* 19 (4): 615–20.

———. 2002. *How Economics Became a Mathematical Science*. Durham: Duke University Press.

———. 2005. "Autobiographical Memory and the Historiography of Economics." *Journal of the History of Economic Thought* 27 (2): 1–11.

———. 2011. "Lionel W. McKenzie and the Proof of the Existence of a Competitive Equilibrium." *Journal of Economic Perspectives* 25 (2): 199–215.

———. 2014. "MIT's Openness to Jewish Economists." In *MIT and the Transformation of American Economics*, ed. E. Roy Weintraub. Durham: Duke University Press.

Weintraub, E. Roy, and Evelyn L. Forget, eds. 2007. *Economists' Lives: Biography and Autobiography in the History of Economics*. History of Political Economy Annual Supplement 39 (5). Durham: Duke University Press.

Weintraub, E. Roy, and Ted Gayer. 2001. "Equilibrium Proofmaking." *Journal of the History of Economic Thought* 23 (4): 421–42.

Westfall, Richard S. 1980. *Never at Rest: A Biography of Isaac Newton*. Cambridge: Cambridge University Press.

White, Hayden. 1987. *The Content of the Form*. Baltimore: Johns Hopkins University Press.

Wittgenstein, Ludwig. 1978. *Remarks on the Foundations of Mathematics*. Rev. ed. Cambridge, MA: MIT Press.

## Archive Material

Kenneth J. Arrow Papers (KJAP). David M. Rubenstein Rare Book and Manuscript Library, Duke University.

Gérard Debreu Papers (GDP Cartons 1–14, additional cartons 1–4). BANC MSS 2006/218, Bancroft Library, University of California, Berkeley.

Frank T. de Vyver Papers (FTDP). David M. Rubenstein Rare Book and Manuscript Library, Duke University.

Oskar Morgenstern Papers (OMP). David M. Rubenstein Rare Book and Manuscript Library, Duke University.

Nicholas Georgescu-Roegen Papers (NGRP). David M. Rubenstein Rare Book and Manuscript Library, Duke University.

Calvin Bryce Hoover Papers (CBHP). David M. Rubenstein Rare Book and Manuscript Library, Duke University.

Lionel W. McKenzie Papers (LWMP). David M. Rubenstein Rare Book and Manuscript Library, Duke University.

Paul A. Samuelson Papers (PASP). David M. Rubenstein Rare Book and Manuscript Library, Duke University.

Robert M. Solow Papers (RMSP). David M. Rubenstein Rare Book and Manuscript Library, Duke University.

# INDEX OF NAMES

# INDEX OF SUBJECTS